Fracture Mechanics

Fracture Mechanics

C. T. Sun

School of Aeronautics and Astronautics
Purdue University
West Lafayette, Indiana

Z.-H. Jin

Department of Mechanical Engineering
The University of Maine
Orono, Maine

AMSTERDAM • BOSTON • HEIDELBERG • LONDON
NEW YORK • OXFORD • PARIS • SAN DIEGO
SAN FRANCISCO • SINGAPORE • SYDNEY • TOKYO

ELSEVIER

Academic Press is an imprint of Elsevier

Academic Press is an imprint of Elsevier
225 Wyman Street, Waltham, MA 02451, USA
The Boulevard, Langford Lane, Kidlington, Oxford, OX5 1GB, UK

Notices

Knowledge and best practice in this field are constantly changing. As new research and experience
broaden our understanding, changes in research methods, professional practices, or medical treatment
may become necessary.

Practitioners and researchers must always rely on their own experience and knowledge in evaluating
and using any information, methods, compounds, or experiments described herein. In using such
information or methods they should be mindful of their own safety and the safety of others, including
parties for whom they have a professional responsibility.

To the fullest extent of the law, neither the Publisher nor the authors, contributors, or editors, assume
any liability for any injury and/or damage to persons or property as a matter of products liability,
negligence or otherwise, or from any use or operation of any methods, products, instructions, or ideas
contained in the material herein.

Library of Congress Cataloging-in-Publication Data
Application submitted.

British Library Cataloguing-in-Publication Data
A catalogue record for this book is available from the British Library.

ISBN: 978-0-12-810337-1

For information on all Academic Press publications
visit our Web site at *www.elsevierdirect.com*

Printed in the United States
11 12 13 14 15   10 9 8 7 6 5 4 3 2 1

Contents

Preface

Fracture mechanics is now considered a mature subject and has become an important course in engineering curricula at many universities. It has also become a useful analysis and design tool to mechanical, structural, and material engineers. Fracture mechanics, especially linear elastic fracture mechanics (LEFM), is a unique field in that its fundamental framework resides in the inverse square root type singular stress field ahead of a crack. Almost all the fracture properties of a solid are characterized using a couple of parameters extracted from these near-tip stress and displacement fields. In view of this unique feature of fracture mechanics, we feel that it is essential for the reader to fully grasp the mathematical details and their representation of the associated physics in these mathematical expressions because the rationale and limitations of this seemingly simple approach are embodied in the singular stress field.

There are already more than a dozen books dealing with fracture mechanics that may be used as textbooks for teaching purposes. With different emphases, these books appeal to different readers and students from different backgrounds. This book is based on the lecture notes that have been used at the School of Aeronautics and Astronautics, Purdue University, for more than 30 years. It is intended as a book for graduate students in aeronautical, civil, mechanical, and materials engineering who are interested in picking up an in-depth understanding of how to utilize fracture mechanics for research, teaching, and engineering applications. As a textbook, our goal is to make it mathematically readable to first-year graduate students with a decent elasticity background. To achieve this goal, almost all mathematical derivations are clearly presented and suitable for classroom teaching and for self-study as well.

In selecting and presenting the contents for this book, we use the aforementioned rationale as a guide. In Chapter 2, the Griffith theory of fracture and the surface energy concept are introduced. Chapter 3 presents the elastic stress and displacement fields near the crack tip and introduces Irwin's stress intensity factor concept. The chapter describes detailed derivations of the stress fields and stress intensity factor K using the complex potential method and Williams' asymptotic expansion approach. Finally, the chapter introduces the fracture criterion based on the stress intensity factor (K-criterion) and discusses the K-dominance concept to make the reader aware of the limitation of the K-criterion.

Chapter 4 is totally devoted to energy release rate in conjunction with the path-independent J-integral. The energy release rate concept is first introduced, and the relationship between the energy release rate G and stress intensity factor ($G - K$ relation) is established followed by the fracture criterion based on the energy release rate (G-criterion). The J-integral is widely accepted because its value is equal to the energy release rate and it can be calculated numerically with stress and displacement fields away from the singular stress at the crack tip. Another simple, yet efficient, crack-closure method has been shown to be quite accurate in evaluating energy release rate. Therefore, a couple of finite element-based numerical methods for calculation of energy release rate and the stress-intensity factor using the crack-closure method are included in this chapter.

In most fracture mechanics books, the near-tip stress field is presented in plane elasticity for Mode I and Mode II loadings and in generalized plane strain for Mode III. In reality, none of these 2-D states exists. For instance, a thin plate containing a center crack is usually treated as a 2-D plane stress problem. In fact, the plane stress assumption fails because of the presence of high stress gradients near the crack tip and a state-of-plane strain actually exists along most part of the crack front. The knowledge of the 3-D nature of all through-the-thickness cracks is important in LEFM. In Chapter 4, a section is devoted to the the 3-D effect on the variation of stress intensity along the crack front.

Under static Mode I loading, experimental results indicate that the direction of crack extension is self-similar. As a result, in determining Mode I fracture toughness of a solid, the crack extension direction is not an issue. The situation is not as clear if the body is subjected to combined loads or dynamic loads. Of course, if the body is an anisotropic solid such as a fiberous composite, the answer to the question of cracking direction is not as simple and is not readily available in general. In view of this constraint, we only consider isotropic brittle solids in Chapter 5. The focus is on the prediction of crack extension direction. From a learning point of view, it is interesting to follow a number of different paths of thinking taken by some earlier researchers in the effort to predict the cracking direction.

Chapters 6 and 7 present the result of the effort in extending the LEFM to treat fracture in elastic-plastic solids. In Chapter 6, plastic zones near the crack tip for the three fracture modes are analyzed. Several popular and simple methods for estimating the crack tip plastic zone size are covered. The initial effort in taking plasticity into account in fractures was proposed by Irwin who suggested using an effective crack length to account for the effect of plasticity. Later, the idea was extended to modeling the so-called R-curve during stable crack growth. Another approach that uses the J-integral derived based on deformation plasticity theory to model the crack tip stress and strain fields (the HRR field) also has many followers. In addition to Irwin's adjusted crack length and the J-integral approach, crack growth modeled by crack tip opening displacement (CTOD or CTOA) is also discussed in Chapter 7.

Interfaces between dissimilar solids are common in modern materials and structures. Interfaces are often the weak link of materials and structures and are the likely sites for crack initiation and propagation. Interfacial cracks have many unique physical behaviors that are not found in homogeneous solids. However, surprisingly, the development of fracture mechanics for interfacial cracks has followed exactly the same path as LEFM. In other words, fracture mechanics for interfacial cracks is all centered on the crack tip stress field. The only difference is in the violently oscillatory behavior of the crack tip stress field of interfacial cracks. Chapter 8 presents a thorough derivation of the crack tip stress and displacement fields. Attention is also focused on the significance of stress oscillation at the crack tip and the nonconvergent nature of the energy release rates of the individual fracture modes.

The cohesive zone model (CZM) has become a popular finite element-based tool for modeling fracture in solids. CZM is often considered by some as a more realistic form of fracture mechanics because it does not employ the idealized singular stresses. Although there are fundamental differences between the two concepts, the purposes of the two are the same. Therefore, it is reasonable to include CZM in this book. In Chapter 9 we make an effort to present the basic formulation of CZM, especially the cohesive traction law. Instead of covering examples of applications of the cohesive zone model, we place greater emphasis on the logic in the formulation of CZM.

Chapter 10 contains brief and condensed presentations of three additional topics, namely, anisotropic solids, nonhomogeneous solids, and dynamic fracture. The reason for including these three topics in this textbook is, perhaps, just for the sake of completeness. For each topic, the coverage is quite brief and with a limited scope and does not warrant a full chapter.

C. T. Sun
Z.-H. Jin

About the Authors

C. T. Sun received his undergraduate education at National Taiwan University. He obtained his M.S. in 1965 and Ph.D. in 1967 from Northwestern University. In 1968 he joined Purdue University, where he is presently Neil A. Armstrong Distinguished Professor in the School of Aeronautics and Astronautics. He has been engaged in composites research for more than forty years. In addition to his work in composites, Professor Sun has published extensively in the areas of fracture mechanics, smart materials, and nanomechanics. He has authored a textbook on aircraft structures published in 1998 with the second edition published in 2006.

Z.-H. Jin is an Associate Professor in the Department of Mechanical Engineering at the University of Maine. He obtained his Ph.D. in Engineering Mechanics from Tsinghua University in 1988. His research areas include fracture mechanics, thermal stresses, mechanical behavior of materials, and geodynamics. He has published more than 70 refereed journal papers and three book chapters.

Fracture Mechanics

Introduction

1.1 FAILURE OF SOLIDS

Failure of solids and structures can take various forms. A structure may fail without breaking the material, such as in elastic buckling. However, failure of the material in a structure surely will lead to failure of the structure. Two general forms of failure in solids are excessive permanent (plastic) deformation and breakage. Plasticity can be viewed as an extension of elasticity for decribing the mechanical behavior of solids beyond yielding. The theory of plasticity has been studied for more than a century and has long been employed for structural designs. On the other hand, the latter form of failure is usually regarded as the strength of a solid, implying the total loss of load-bearing capability of the solid. For brittle solids, this form of failure often causes the body under load to break into two or more separated parts.

Unlike plasticity, the prediction of the strength of solid materials was all based on phenomenological approaches before the inception of fracture mechanics. Many phenomenological failure criteria in terms of stress or strain have been proposed and calibrated against experimental results. In the commonly used failure criteria, such as the maximum principal stress or strain criterion, a failure envelope in the stress or strain space is constructed based on limited experimental strength data. Failure is assumed to occur when the maximum normal stress at a point in the material exceeds the strength envelope, that is,

$$\sigma_1 \geq \sigma_f$$

where σ_1 (> 0) is a principal stress and σ_f is the tensile strength of the solid. The failure envelope has also been modified to distinguish the difference between tensile and compressive strengths and to account for the effects of stress interactions.

In general, the classical phenomenological failure theories predict failure of engineering materials and structures with reasonable accuracy in applications where the stress field is relatively uniform. These theories are often unreliable in the presence of high-stress gradients resulting from cutouts. Moreover, there were many premature structural failures at stresses that were well below the critical values specified in the classical failure theories.

Fracture Mechanics

1

The most frequently cited example is the failure of Liberty cargo ships built during World War II. Among roughly 2700 all-welded hull ships, more than 100 were seriously fractured and about 10 were fractured in half [1-1]. It was demonstrated [1-2] that cracks were first initiated at the stress concentration locations and then propagated in the hull, resulting in the catastrophic failure. Other significant examples include fuselage failure in Comet passenger jet airplanes from 1953 to 1955 [1-3] and failure of heavy rotors in steam turbines from 1955 to 1956 [1-4].

The aforementioned historical events led researchers to recognize that defects are the original cause of failure and in strength predictions, materials cannot be always assumed free of defects. Cracks and other forms of defects may be introduced during materials manufacturing and processing, as well as during service. For instance, rapid quenching of cast irons results in microcracks in the material. Cyclic stresses induce cracks in the connections of the structural components. The stresses at the crack tip are much higher than the material strength, which is measured under a state of uniform stress in laboratory condition. The high stresses near the crack tip drive the crack to extend, leading to the eventual catastrophic failure of the material. Failure caused by crack propagation is usually called *fracture failure*. The classical failure criteria assume that materials are free of defects, and hence are not capable of predicting fracture failure, or failure of materials containing crack-like flaws.

1.2 FRACTURE MECHANICS CONCEPTS

Fracture mechanics is a subject of engineering science that deals with failure of solids caused by crack initiation and propagation. There are two basic approaches to establish fracture criteria, or crack propagation criteria: crack tip stress field (local) and energy balance (global) approaches. In the crack tip field approach, the crack tip stress and displacement states are first analyzed and parameters governing the near-tip stress and displacement fields are identified. Linear elastic analysis of a cracked body shows that stresses around the crack tip vary according to $r^{-1/2}$, where r is the distance from the tip. It is clear that stresses become unbounded as r approaches the crack tip. Such a singular stress field makes the classical strength of materials failure criteria inapplicable.

A fundamental concept of fracture mechanics is to accept the theoretical stress singularity at the crack tip but not use the stress directly to determine failure/crack extension. This is based on the fact that the tip stress is limited by the yield stress or the cohesive stress between atoms and singular stresses are the results of linear elasticity. It is also recognized that the singular stress field is a convenient representation of the actual finite stress field if the discrepancy between the two lies in a small region near the crack tip. This notion is referred to as small-scale yielding.

The stresses near the tip of a crack in linearly elastic solids have the following universal form independent of applied loads and the geometry of the cracked body

(Chapter 3):

$$\sigma_{xx} = \frac{K_I}{\sqrt{2\pi r}} \cos\frac{1}{2}\theta \left(1 - \sin\frac{1}{2}\theta \sin\frac{3}{2}\theta\right)$$

$$\sigma_{yy} = \frac{K_I}{\sqrt{2\pi r}} \cos\frac{1}{2}\theta \left(1 + \sin\frac{1}{2}\theta \sin\frac{3}{2}\theta\right) \tag{1.1}$$

$$\sigma_{xy} = \frac{K_I}{\sqrt{2\pi r}} \sin\frac{1}{2}\theta \cos\frac{1}{2}\theta \cos\frac{3}{2}\theta$$

where K_I is the so-called stress intensity factor, which depends on the applied load and crack geometry and (r,θ) are the polar coordinates centered at the crack tip. Here it is assumed that the loads and the geometry are symmetric about the crack line. Equation (1.1) shows that K_I is a measure of the stress intensity near the crack tip.

Based on this obervation, Irwin [1-5] proposed a fracture criterion which states that crack growth occurs when the stress intensity factor reaches a critical value, that is,

$$K_I = K_{Ic} \tag{1.2}$$

where K_{Ic} is called fracture toughness, a material constant determined by experiment. The preceding fracture criterion for cracked solids is fundamentally different from the classical failure criteria based on stresses. It does not directly use stresses or strains, but a proportionality factor in the stress field around the crack tip. K_I is proportional to the applied load but has a dimension of $MPa - \sqrt{m}$ in the SI unit system and $ksi - \sqrt{in}$ in the US customary unit system. K_{Ic} is a new material parameter introduced in fracture mechanics that characterizes the resistance of a material to crack extension.

The criterion in Eq. (1.2) is based on linear elasticity with which the inverse square root singular stress field exists and the stress intensity factor is well defined. The actual fracture process at the crack tip cannot be described using the linear elasticity theory. The rationality of the criterion lies in the condition that the fracture process zone is sufficiently small so that it is well contained inside the singular stress field Eq. (1.1) characterized by the stress intensity factor K_I.

The second approach for establishing a fracture criterion is based on the consideration of global energy balance during crack extension. The potential energy of a cracked solid under a given load is first determined and its variation with a virtual crack extension is then examined. Consider a two-dimensional elastic body with a crack of length a. The total potential energy per unit thickness of the system is denoted by $\Pi = \Pi(a)$. Note that the potential energy is a function of the crack length. For a small crack extension da, the decrease in the potential energy is $-d\Pi$. Griffith [1-6] proposed that this energy decrease in the cracked body would be absorbed into the surface energy of the newly created crack surface. Denote the surface energy per unit area by γ, which can be calculated from solid state physics. The total surface energy of the new crack surface equals $2da\gamma$. The Griffith energy balance equation becomes

$$-d\Pi = 2da\gamma \quad \text{or} \quad -\frac{d\Pi}{da} = 2\gamma$$

The energy release rate G proposed by Irwin [1-7] is defined as the decrease in potential energy per unit crack extension under constant load, that is,

$$G = -\frac{d\Pi}{da}$$

The crack growth or failure criterion using the energy balance approach is established as

$$G = G_c = 2\gamma \tag{1.3}$$

The fracture criterion given before is also fundamentally different from the classical failure criteria. It involves the total energy of the cracked body as well as the surface energy of the solid, which exists only in atomistic scale considerations. Like K_{Ic}, G_{Ic} is also a new material constant introduced in fracture mechanics to measure the resistance to fracture. G_{Ic} has a dimension of J/m^2, or kJ/m^2. The fracture criteria Eqs. (1.2) and (1.3) are actually equivalent (Chapter 4). However, the experimentally measured critical energy release rate for engineering materials, especially metals, is significantly larger than 2γ. This is because plastic deformations in the crack tip region also contribute significantly to the crack growth resistance. For perfectly brittle solids, it has been shown by MD simulations that $G_{Ic} = 2\gamma$ is valid in NaCl single crystal if the crack length is equal to or greater than 10 times the lattice constant [1-8].

Fracture mechanics introduces two novel concepts: stress intensity factor and energy release rate. These two quantities distinguish fracture mechanics from the classical failure criteria. In using the stress intensity factor-based fracture criterion to predict failure of a material or structure, one first needs to calculate the stress intensity factor for the given load and geometry. The second step is to measure the fracture toughness. Once the stress intensity factor and the fracture toughness are known, Eq. (1.2) can be used to determine the maximum allowable load that will not cause crack growth for a given crack length, or the maximum allowable crack length that will not propagate under the design load. The advantage of the stress intensity factor approach is its ease in the calculation of stress intensity factors and the easy measurement of fracture toughness. In contrast to the stress intensity factor approach, the energy release rate-based fracture criterion Eq. (1.3) is more naturally extended to cases where nonlinear effects need to be accounted for because the energy concept is universal.

Stress intensity factor and energy release rate lay the foundation of linear elastic fracture mechanics (LEFM). In LEFM, the cracked solid is treated as a linearly elastic medium and nonlinear effects are assumed to be minimal and can be ignored. While a modified stress intensity factor approach may be used to predict fracture of a cracked solid when plastic deformations are small and confined in the near-tip region, the approach, along with the energy release rate, would become futile when the cracked solid undergoes large-scale plastic deformations. Several fracture parameters have been proposed to predict fracture of solids under nonlinear deformation conditions, for example, the J-integral, the crack tip opening displacement (CTOD), and the crack tip opening angle (CTOA). Failure criteria based on these parameters, however, have not been as successful as the stress intensity factor and the energy release rate critetia in LEFM.

1.3 HISTORY OF FRACTURE MECHANICS

This section briefly describes the historical development of fracture mechanics from Griffith's pioneering work on brittle fracture of glass in 1920s, to Irwin's stress intensity factor concept and fracture criterion in 1950s, and to elastic-plastic fracture mechanics research in 1960s and early 1970s. A brief introduction of recent development of fracture mechanics research since 1990s is also included.

1.3.1 Griffith Theory of Fracture

The advent of fracture mechanics is usually credited to the poineering work of A. A. Griffith on brittle fracture of glass [1-6]. This paper was basically his PhD thesis work at Cambrige University under the guidance of G. I. Taylor. It had been known before Griffith's work that the theoretical fracture strength of glass determined based on the breaking of atomic bonds exceeds the strength of laboratory specimens by one to two orders of magnitude. Griffith believed that this huge discrepancy could be due to microcracks in the glass and that these cracks could propagate under a load level that is much smaller than the theoretical strength.

Griffith adopted an energy balance approach to determine the strength of cracked solids, that is, the work done during a crack extension must be equal to the surface energy stored in the newly created surfaces. To calculate the strain energy in a cracked body, he derived the stress field in an infinite plate with a through-thickness central crack under biaxial loading from Inglis's solution [1-9] for an elliptical hole in an elastic plate by reducing the minor axis to zero. Using this solution, Griffith was able to calculate the total potential energies before and after crack extension. The difference of the potential enegies of the two states were set equal to the corresponding gain in surface energy.

It follows from the Griffith theory that the fracture strength (the remote applied stress) of a solid with a crack is proportional to the square root of the surface energy and is inversely proportional to the square root of the crack size, that is,

$$\sigma_f \propto \sqrt{\frac{\gamma_c E}{a}}$$

where σ_f is the applied failure stress, γ_c is the specific surface energy, a is half the crack length, and E is Young's modulus.

The preceding relationship points out a specific functional form between the failure stress and the crack size. The Griffith theory represents a breakthrough in the strength theory of solids. It successfully explains why there is an order of magnitude difference between the theoretical strength and experimentally measured failure load for a solid. In particular, it provides a well-defined physical mechanism that controls the failure process, which is lacking in the classical phenomenological failure theories. The original work of A. A. Griffith dealt with fracture of brittle glass. In metals, plastic deformations develop around the crack tip and the measured fracture strength is much greater than that predicted by the Griffith theory. Orowan [1-10] and Irwin [1-11] suggested to add to 2γ the plastic work γ_p associated with the creation of new crack surfaces. For metals, γ_p is much larger than the surface energy 2γ, and hence the modified Griffith theory by Orowan and Irwin explained the high fracture strength of metals.

1.3.2 Fracture Mechanics as an Engineering Science

Although the basic energy concept of fracture mechanics was presented by A. A. Griffith in 1920, it was only after the 1950s that fracture mechanics was accepted as an engineering science with successful practical applications mainly as a result of Irwin's work ([1-5] and [1-7]). Irwin first introduced the energy release rate to establish a fracture criterion as in Eq. (1.3). He then defined the stress intensity factor K and derived the relationship between the energy release rate G and the stress intensity factor K based on Westergaard's solutions for the stress and displacement fields in a cracked plate [1-12]. Because of the $G - K$ relationship, Irwin proposed to use the stress intensity factor as a fracture parameter, which is a more direct approach for fracture mechanics applications as described by Eq. (1.2).

At the same time, Williams [1-13] derived the asymptotic stress field near a crack tip with the leading term exhibiting an inverse square root singularity under general planar loading conditions. The Williams solution, with both symmetric and asymmetric terms, gives a universal expression for the crack tip stress field independent of external loads and crack geometries. The load and crack geometry influence the crack tip singular stresses through the stress intensity factors K_I and K_{II}, which govern the intensity of the singular stress field. Williams' solution provides a justification for adopting the stress intensity factors to establish fracture criteria.

The stress intensity factor fracture criterion assumes that materials behavior is linearly elastic, which is a good assumption for brittle materials such as glass and ceramics. For ductile metals at room and elevated temperatures, however, plastic yielding occurs around the crack tip due to the stress singularity predicted in the elastic solution. For linear elastic fracture mechanics to be applicable to metals, the plastic deformation zone around the crack tip must be smaller than the dominance zone of the stress intensity factor. Irwin [1-14] estimated the size of plastic deformation zone near the crack tip and found that the plastic zone size is proportional to the square of the stress intensity factor to the yield strength ratio if the plastic zone is small.

With the fracture criterion Eq. (1.2) in hand and the knowledge of the crack tip plastic zone size, the American Society for Testing and Materials (ASTM) formed a Special Technical Committee (ASMT STC, subsequently ASTM Committee E-24) to develop the standard for measuring K_{Ic}, the plane strain fracture toughness (or simply fracture toughness) for metallic materials. In the meantime, great efforts were made in 1960s and 1970s to develop analytical and numerical methods to compute stress intensity factors. Most of the approaches and techniques are included in a multi-volume fracture mechanics monograph, *Mechanics of Fracture*, edited by G. C. Sih and his coworkers [1-15, 1-17]. Stress intensity factors for various crack geometries under a variety of loading conditions are compiled in the handbook by Tada et al. [1-18].

The LEFM based on stress intensity factor K and energy release rate G has been very successful in predicting fracture of metals when the crack tip plastic zone is smaller than the K-dominance zone—also termed small-scale yielding (SSY). Under large-scale yielding conditions, however, the LEFM generally becomes inadequate and fracture criteria based on plasticity of the cracked solids have to be used. Irwin [1-14] introduced an effective stress intensity factor concept to take the crack tip plasticity effect into account. The effective stress intensity factor is obtained by replacing the crack length with an effective crack length that is equal to the original length plus half the plastic zone size.

Dugdale [1-19] presented a strip yielding zone model to determine the plastic zone size in thin cracked sheets. Wells [1-20] and [1-21] proposed to use the crack opening dispalcement (COD) as a fracture parameter. The COD criterion is equivalent to the effective stress intensity factor criterion under modest yielding conditions but can be extended to large-scale yielding when it is combined with the COD equation from the Dugdale model. Rice [1-22] generalized the energy release rate concept to nonlinear elastic materials or elastic-plastic materials described by the deformation plasticity and found that the energy release rate can be represented by a line integral, the so-called path-independent J-integral.

Begley and Landes [1-23] later proposed to use the J-integral for predicting elastic-plastic crack initiation and experimentally measured the critical value of J at crack initiation. In 1968, Rice and Rosengren [1-24] and Hutchinson [1-25] published their work on the crack tip plastic stress field (HRR field) in the framework of deformation plasticity. The HRR field shows that the J-integral characterizes the intensity of the singular stress field in a similar way to the role of stress intensity factor in LEFM. Becasue the HRR field is based on the deformation plasticity, the J-integral in general may be used for crack initiation only. In other words, the HRR field disappears as the crack extends and unloading (a behavior that the deformation plasticity theory cannot model) takes place.

1.3.3 Recent Developments in Fracture Mechanics Research

In recent years, LEFM has found many new applications mostly dealing with new materials such as nonhomogeneous and anisotropic fiber-reinforced composites.

The main issues that arise in these new applications include, for example, coupled thermal-mechanical loads in microelectronic packaging and multiscale issues in treating composite materials as homogeneous solids. Because of the increasing interest in nanotechnology, fracture of nanostructured materials has recently attracted the attention of many researchers. Molecular dynamics (MD) simulations are employed to model crack extension in atomistic systems. Researchers have attempted to answer the question regarding the applicability of continuum theory-based LEFM in solids at nano scale ([1-8] and [1-26]). For instance, are the stress intensity factor and energy release rate introduced in LEFM still valid, and how can one evaluate their values? Other issues involve the definition of cracks that are equivalent to cracks adopted in continuum LEFM.

A new form of fracture model called cohesive zone model (CZM) has evolved from LEFM but has taken a different treatment of the crack tip stress and strain fields. The main motivation in CZM was to avoid the seemingly unrealistic stress singularity at the crack tip. The idea of the CZM is credited to Barenblatt [1-27], who assumed that failure would occur by decohesion of the upper and lower surfaces of a volumeless cohesive zone ahead of the crack tip. In the cohesive zone the separation displacement of the two surfaces that bound the cohesive zone follows a cohesive traction law. Crack growth occurs when the opening displacement at the tail of the cohesive zone (physical crack tip) reaches a critical value at which the cohesive traction vanishes. Clearly, the cohesive modeling approach does not involve stress singularities and material failure is controlled by quantities such as displacements and stresses, which are consistent with the usual strength of materials theory.

Since Needleman [1-28] introduced the cohesive element technique in the finite element framework for fracture studies, CZM has emerged as a popular tool for simulating fracture processes in materials and structures due to the computational convenience. Although many researchers have reported successful results using the CZM approach, many issues remain to be resolved including the physics of the cohesive zone, a rational way to develop the cohesive traction law, and the uniqueness of the cohesive traction with respect to variation of loads and specimen geometry.

References

[1-1] H.P. Rossmanith, The struggle for recognition of engineering fracture mechanics, in: H.P. Rossmanith (Ed.), Fracture Research in Retrospect, A.A. Balkema, Rotterdam, Netherlands, 1997, pp. 37–94.

[1-2] E. Hayes, Dr. Constance Tipper: testing her mettle in a materials world, Adv. Mater. Processes Vol. 153, 100. (1998).

[1-3] A.A. Wells, The condition of fast fracture in aluminum alloys with particularly reference to Comet failures, British Welding Research Association Report, NRB 129, April 1955.

[1-4] D.J. Winne, B.M. Wundt, Application of the Griffith-Irwin theory of crack propagation to the bursting behavior of disks, including analytical and experimental studies, Trans. ASME 80 (1958) 1643–1655.

[1-5] G.R. Irwin, Analysis of stresses and strains near the end of a crack traversing a plate, J. Appl. Mech. 24 (1957) 361–364.

[1-6] A.A. Griffith, The phenomena of rapture and flow in solids, Philos Trans R Soc Lond A221 (1920) 163–198.

[1-7] G.R. Irwin, Relation of stresses near a crack to the crack extension force, in: Proceedings of the International Congresses of Applied Mechanics, Vol. VIII, University of Brussels, 1957, pp. 245–251.

[1-8] A. Adnan, C.T. Sun, Evolution of nanoscale defects to planar cracks in a brittle solid, J. Mech. Phys. Sol. 58 (2010) 983–1000.

[1-9] C.E. Inglis, Stresses in a plate due to the presence of cracks and sharp corners, Trans. Inst. Naval Architects 55 (1913) 219–230.

[1-10] E. Orowan, Notch brittleness and strength of metals, Trans. Inst. Engrs Shipbuilders in Scotland 89 (1945) 165–215.

[1-11] G.R. Irwin, Fracture dynamics, in: Fracture of Metals, ASM, Cleveland, OH, 1948, pp.147–166.

[1-12] H.M. Westergaard, Bearing pressures and cracks, J. Appl. Mech. 6 (1939) 49–53.

[1-13] M.L. Williams, On the stress distribuion at the base of a stationary crack, J. Appl. Mech., 24 (1957) 109–114.

[1-14] G.R. Irwin, Plastic zone near a crack and fracture toughness, in: Proceedings of the 7th Sagamore Ordnance Materials Conference, Syracuse University, 1960, pp. IV-63-IV-78.

[1-15] G.C. Sih (Ed.), Mechanics of Fracture Vol. 1: Methods of Analysis and Solutions of Crack Problems, Leyden, Noordhoff International Pub., 1973.

[1-16] M.K. Kassir, G.C. Sih (Eds.), Mechanics of Fracture Vol. 2: Three-Dimensional Crack Problems, Leyden, Noordhoff International Pub., 1975.

[1-17] G.C. Sih (Ed.), Mechanics of Fracture Vol. 3: Plates and Shells with Cracks, Leyden, Noordhoff International Pub., 1977.

[1-18] H. Tada, P.C. Paris, G.R. Irwin, The Stress Analysis of Cracks Handbook, ASME Press, New York, 2000.

[1-19] D.S. Dugdale, Yielding of steel sheets containing slits, J. Mech. Phys. Sol. 8 (1960) 100–104.

[1-20] A.A. Wells, Unstable crack propagation in metals: cleavage and fast fracture, in: Proceedings of the Crack Propagation Symposium, Vol. 1, Paper 84, Cracnfield, UK, 1961.

[1-21] A.A. Wells, Application of fracture mechanics at and beyond general yielding, Br. Weld. J. 11 (1963) 563–570.

[1-22] J.R. Rice, A path independent integral and the approximate analysis of strain concentration by notches and cracks, ASME J. Appl. Mech. 35 (1968) 379–386.

[1-23] J.A. Begley, J.D. Landes, The J-integral as a fracture criterion, in: ASTM STP 514, American Society for Testing and Materials, Philadelphia, 1972, pp. 1–20.

[1-24] J.R. Rice, G.F. Rosengren, Plane strain deformation near a crack tip in a power-law hardening material, J. Mech. Phys. Sol. 16 (1968) 1–12.

[1-25] J.W. Hutchinson, Singular behavior at the end of a tensile crack in a hardening material, J. Mech. Phys. Sol. 16 (1968) 13–31.

[1-26] M.J. Buehler, H. Yao, B. Ji, H. Gao, Cracking and adhesion at small scales: atomistic and continuum studies of flaw tolerant nanostructures, Model. Simul. Mater. Sci. Eng. 14 (2006) 799–816.

[1-27] G.I. Barenblatt, The mathematical theory of equilibrium cracks in brittle fracture, in: Adv. Appl. Mech. 7 (1962) 55–129.

[1-28] A. Needleman, A continuum model for void nucleation by inclusion debonding, J. Appl. Mech. 54 (1987) 525–531.

Griffith Theory of Fracture

2.1 THEORETICAL STRENGTH

The theoretical strength of a solid is usually understood as the applied stress that fractures a perfect crystal of the material by breaking the atomic bonds along the fractured surfaces. The theoretical strength may be estimated using the interatomic bonding force versus the atomic separation relation. This section gives two estimates that relate the theoretical strength to the Young's modulus of the material based on the atomic bonding strength and surface energy concept.

2.1.1 An Atomistic Model

In general, failure of a solid is characterized by separation of the body. At the atomistic level the fracture strength of a "perfect" material depends on the strength of its atomic bonds. Consider two arrays of atoms in a perfect crystal as shown in Figure 2.1. Let a_0 be the equilibrium spacing between atomic planes in the absence of applied stresses. The stress σ required to separate the planes to a distance $a > a_0$ increases until the theoretical strength σ_c is reached and the bonds are

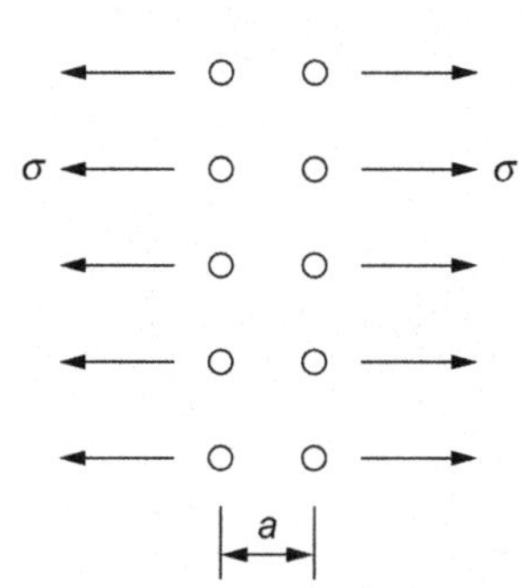

FIGURE 2.1

Atomic planes in a perfect crystal.

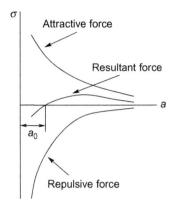

FIGURE 2.2

Cohesive force versus separation between atoms.

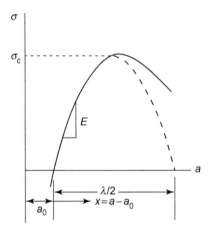

FIGURE 2.3

Cohesive force between atoms.

broken. Further displacements of the atoms can occur under a decreasing applied stress (see Figure 2.2). This stress-displacement curve can be approximated by a sine curve (Figure 2.3) having wavelength λ as

$$\sigma = \sigma_c \sin\left(\frac{2\pi x}{\lambda}\right) \qquad (2.1)$$

where $x = a - a_0$ is the relative displacement between the atoms.

At small displacement x we have

$$\sin x \simeq x$$

and, thus,

$$\sigma \simeq \sigma_c \frac{2\pi x}{\lambda} \qquad (2.2)$$

The modulus of elasticity is

$$E = \frac{\text{stress}}{\text{strain}} = \frac{\sigma}{x/a_0} \Rightarrow \sigma = \frac{Ex}{a_o} \qquad (2.3)$$

Using Eqs. (2.2) and (2.3), we obtain

$$\frac{Ex}{a_0} = \sigma_c \frac{2\pi x}{\lambda}$$

or

$$\sigma_c = \frac{\lambda E}{2\pi a_0} \qquad (2.4)$$

A reasonable value for λ is $\lambda = a_o$, which yields the bond strength

$$\sigma_c = \frac{E}{2\pi} \qquad (2.5)$$

2.1.2 The Energy Consideration

Theoretical strength may also be estimated using the simple atomic model with a surface energy concept. We now define a quantity called the surface energy γ (energy per unit area) as the work done in creating new surface area by the breaking of atomic bonds. From the sine-curve approximation of the atomic force (see Figure 2.3), this is simply one-half the area under the stress-displacement curve since two new surfaces are created each time a bond is broken. Thus,

$$2\gamma = \int_0^{\lambda/2} \sigma_c \sin\left(\frac{2\pi x}{\lambda}\right) dx = \frac{\lambda \sigma_c}{\pi}$$

from which

$$\sigma_c = \frac{2\gamma \pi}{\lambda} \qquad (2.6)$$

However, from Eq. (2.4),

$$\lambda = \frac{2\pi a_0 \sigma_c}{E} \qquad (2.7)$$

We obtain from substitution of Eq. (2.7) in Eq. (2.6)

$$\sigma_c^2 = \frac{2\pi \gamma E}{2\pi a_0} = \frac{\gamma E}{a_0}$$

Finally,

$$\sigma_c = \sqrt{\frac{\gamma E}{a_0}} \tag{2.8}$$

For many materials, γ is on the order of $0.01 E a_0$ [2-1]. Thus, an approximate estimation of the theoretical strength is often given by

$$\sigma_c = \frac{E}{10} \tag{2.9}$$

which agrees with Eq. (2.5) in terms of order of magnitude. For most metals, the theoretical strength varies between 7 GPa (1×10^6 psi) and 21 GPa (3×10^6 psi). However, bulk materials that are commercially produced for engineering applications commonly fracture at applied stress levels 10 to 100 times below these values, and the theoretical strength is rarely obtained in engineering practice.

The main reasons for the discrepancies are

1. The existence of stress concentrators (flaws such as cracks and notches)
2. The existence of planes of weakness such as grain boundaries in polycrystalline materials.

Another type of fracture process (e.g., shear or rupture), which occurs by plastic deformation, intervenes at a lower level of applied stress.

2.2 THE GRIFFITH THEORY OF FRACTURE

Alan Griffith's work [2-2] on brittle fracture of glass was motivated by the desire to explain the discrepancy between the theoretical strength and actual stength of materials. According to the preceding theoretical strength calculation, we may conclude that glass should be very strong. However, laboratory test results often indicate otherwise. Griffith argued that what we must account for is not the strength but rather the weakness, which is normally dominant in the failure process. One clue obviously lies in the fact that actual glasses display a far more complex fracture behavior than predicted by our simple assumptions regarding the cohesive strength. In his pioneer paper [2-2], Griffith postulated that all bulk glasses contain numerous minute flaws in the form of microcracks that act as stress concentration generators. This new concept accompanied by the energy release approach that he introduced started the era of modern fracture mechanics.

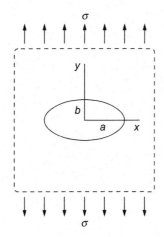

FIGURE 2.4

An elliptic hole in an infinite plate subjected to tension.

The solution of an elliptic hole in a plate of infinite extent under tension by Inglis [2-3] was the first step toward relating observed failure stress to ultimate strength. He solved the problem as shown in Figure 2.4 and found that the greatest stress occurs at the ends of the major axis:

$$\sigma_{yy} = \sigma \left(1 + \frac{2a}{b} \right) \tag{2.10}$$

If $a = b$ (a circular hole), then $\sigma_{yy} = 3\sigma$, which yields the well-known stress concentration factor near a circular hole. If $b \to 0$, then we have a "line crack," and the stress σ_{yy} increases without limit. If a stress-based failure criterion is used to predict the extension of such a "crack," one would find the unreasonable answer that any amount of applied stress would cause the crack to grow.

Griffith took an energy balance point of view and reasoned that the unstable propagation of a crack must result in a decrease in the strain energy of the system (for a body with a fixed boundary where no work is done by external forces during the crack extension), and proposed that a crack would advance when the incremental release of energy dW associated with a crack extension da in a body becomes greater than the incremental increase of surface energy dW_S as new crack surfaces are created. That is,

$$dW \geq dW_S \tag{2.11}$$

The equality indicates the critical point for crack propagation. In other words, if the supply of energy from the cracked plate is equal to or greater than the energy required to create new crack surfaces, the crack can extend.

It is easy to calculate the surface energy for a crack (having two crack tips) with length $2a$, that is,

$$W_S = 2(2a\gamma) = 4a\gamma \tag{2.12}$$

in which γ is the surface energy density and the fact that two crack surfaces for a crack has been accounted for.

Griffith used Inglis' solution to obtain the total energy released W due to the presence of a crack of length $2a$ in an infinite two-dimensional body. His method for calculating energy release was very complicated since he considered the energy change in the body as a whole. He obtained, for plane strain,

$$W = \frac{\pi a^2 \sigma^2 (1 - v^2)}{E} \tag{2.13}$$

and for plane stress,

$$W = \frac{\pi a^2 \sigma^2}{E} \tag{2.14}$$

Thus, the critical stress σ_{cr} under which the crack would start propagating may be obtained by substituting Eqs. (2.12) and (2.13) or Eq. (2.14) into Eq. (2.11):

$$\frac{2\pi a(1 - v^2)\sigma_{cr}^2}{E} da = 4\gamma \, da \qquad \text{(plane strain)}$$

From this equation, we have

$$\sigma_{cr}^2 = \frac{4\gamma E}{2\pi a(1 - v^2)}$$

or

$$\sigma_{cr} = \sqrt{\frac{2E\gamma}{\pi(1 - v^2)a}} \tag{2.15}$$

Similarly, for plane stress we have

$$\sigma_{cr} = \sqrt{\frac{2E\gamma}{\pi a}} \tag{2.16}$$

Comparing the previous critical stress σ_{cr}, or the fracture strength of the infinite plate with a crack of microscopic or macroscopic length $2a$, and the theoretical strength σ_c in Eq. (2.8), we note $\sigma_c \gg \sigma_{cr}$ if $a \gg a_o$. This explains qualitatively why actual strengths of materials are much smaller than their theoretical strengths.

The energy released dW for a crack extension of da can be expressed in terms of the "strain energy release rate per crack tip" G as

$$dW = 2G \, da \tag{2.17}$$

Thus,

$$G = \frac{1}{2}\frac{dW}{da} = \frac{\pi a \sigma^2 (1 - \nu^2)}{E} \quad \text{for plane strain} \tag{2.18}$$

$$= \frac{\pi a \sigma^2}{E} \quad \text{for plane stress} \tag{2.19}$$

The instability condition then reads

$$G \geq 2\gamma \tag{2.20}$$

The value of G when equal to 2γ is denoted by G_c and is called the **fracture toughness** or the **crack-resistant force** of the material. This is, in fact, the Griffith energy criterion of brittle fracture. In theory, a crack would extend in a brittle material when the load produces an energy release rate G equal to 2γ. However, such an energy release rate turns out to be much smaller than the test data since most materals are not perfectly brittle and plastic deformation occurs near the crack tip.

To take this additional crack resistant force into account, Orowan [2-4] suggested to add to 2γ the plastic work γ_p associated with the creation of the new crack surfaces. For metals, 2γ is usually much smaller than γ_p and, thus, can be neglected. On the other hand, Irwin [2-5] took G_c as a new material constant to be measured directly from fracture tests. However, Eq. (2.15) points to a correct relation between the failure stress and crack size.

2.3 A RELATION AMONG ENERGIES

The Griffith theory for fracture of perfectly brittle elastic solids is founded on the principle of energy conservation that is, energy added to and released from the body must be the same as that dissipated during crack extension. It states that, during crack extension of da, the work done dW_e by external forces, the increment of surface energy dW_S, and the increment of elastic strain energy dU must satisfy

$$dW_S + dU = dW_e \tag{2.21}$$

For a conservative force field, this condition can be expressed in the form

$$\partial(W_S + U + V)/\partial a = 0 \tag{2.22}$$

where

$W_S =$ total crack surface energy associated with the entire crack

$U =$ total elastic strain energy of the cracked body

$V =$ total potential of the external forces

Note that a negative dV implies a positive work dW_e done by external forces.

Consider a single-edge-cracked elastic specimen subjected to a tensile load P or displacement δ as shown in Figure 2.5. The relationship between the applied tensile

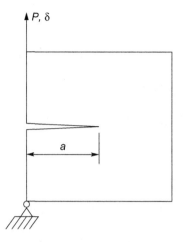

FIGURE 2.5

A single-edge-cracked specimen.

force P and the elastic extension, or displacement, δ, is

$$\delta = SP \tag{2.23}$$

where S denotes the elastic compliance of the specimen containing the crack. The strain energy stored in this specimen is

$$U = \int_{\delta=0}^{\delta=SP} P d\delta = \int_{\delta=0}^{\delta=SP} \frac{\delta}{S} d\delta$$

$$= \frac{1}{2S}[\delta^2]_0^{SP} = \frac{1}{2}SP^2 \tag{2.24}$$

The compliance S is a function of the crack length. The incremental strain energy under the condition of varying a and P is

$$dU = \frac{1}{2}P^2 dS + SP dP \tag{2.25}$$

Case 2.1

Suppose that the boundary is fixed during the extension of the crack so that

$$\delta = SP = \text{constant}$$

Consequently,

$$d\delta = SdP + PdS = 0$$

from which we obtain

$$SdP = -PdS$$

Substitution of the preceding equation into Eq. (2.25) yields

$$dU|_\delta = -\frac{1}{2}P^2 dS \qquad (2.26)$$

Furthermore, $dW_e = 0$ in this case because $d\delta = 0$ and, thus, the external load does no work. Substituting Eq. (2.26) into Eq. (2.21) and using $dW_e = 0$, we have

$$dW_S = -dU|_\delta = \frac{1}{2}P^2 dS \qquad (2.27)$$

Thus, a decrease in strain energy U is compensated by an increase of the same amount in the surface energy. In other words, the energy consumed during crack extension is entirely supplied by the strain energy stored in the cracked body.

Case 2.2

Suppose that the applied force is kept constant during crack extension; then

$$dP = 0$$

From Eq. (2.25) we have

$$dU|_P = \frac{1}{2}P^2 dS \qquad (2.28)$$

Thus, there is a gain in strain energy during crack extension in this case. Moreover, we note that

$$dW_e = Pd\delta = P^2 dS \qquad (2.29)$$

Substituting Eqs. (2.28) and (2.29) into Eq. (2.21), we again obtain Eq. (2.27), that is,

$$dW_S = \frac{1}{2}P^2 dS$$

which is half of the work done by the external force. It is interesting to note that the work done by the external force is split equally into the surface energy and an increase in strain energy.

For both boundary conditions discussed before, the energy released during crack extension is

$$dW = dW_e - dU = \frac{1}{2}P^2 dS$$

The corresponding energy release rate is

$$G = \frac{dW}{da} = \frac{1}{2}P^2\frac{dS}{da}$$

(2.30)

Hence, the strain energy release rate is independent of the type of loading.

The two loading cases can be illustrated graphically as in Figures 2.6a and 2.6b, respectively. In the figures, point B indicates the beginning of crack extension and point C the termination of crack extension. The area $\overline{OBC}$ is the strain energy released, dW. It can be shown rather easily from the graphic illustration that the energies released in the two cases are equal.

Under the fixed load condition, we have

$$dW = dU = \frac{1}{2}dW_e$$

Thus, the energy release rate can be obtained with

$$G = \frac{dU}{da}$$

(2.31)

in which the differentiation is performed assuming that the applied load is independent of a.

Under the fixed displacement condition, we have $dW = -dU$, and hence

$$G = -\frac{dU}{da}$$

(2.32)

In the previous equation, the applied load P should be considered as a function of crack length a in the differentiation. The result should be the same as that given by Eq. (2.31). It is noted that the relation $dW = dU = dW_e/2$ is not true for nonlinear solids.

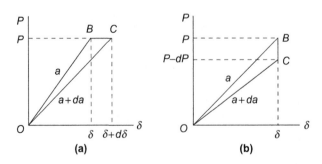

FIGURE 2.6

Energy released during crack extension: (a) constant load, (b) constant displacement.

Example 2.1

The double cantilever beam (DCB) is often used for measuring fracture toughness of materials. Consider the geometry shown in Figure 2.7 where b is the width of the beam, and the crack length a is much larger than h and, thus, the simple beam theory is suitable for modeling the deflection of the two split beams.

Noting that the unsplit portion of the DCB is not subjected to any load and that in each leg the bending moment is $M = Px$, we calculate the total strain energy stored in the two legs of the DCB as

$$U_T = 2 \int_0^a \frac{P^2 x^2}{2EI} dx = \frac{P^2 a^3}{3EI}$$

where

$$I = \frac{bh^3}{12}$$

The total strain energy per unit width is

$$U = U_T / b$$

The strain energy release rate is obtained as

$$G = \frac{dU}{da} = \frac{P^2 a^2}{bEI}$$

If the fracture toughness G_c of the material is known, then the load that could further split the beam is

$$P_{cr} = \frac{\sqrt{bEIG_c}}{a}$$

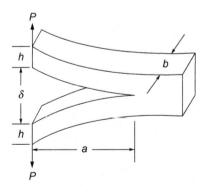

FIGURE 2.7

A double cantilever beam subjected to concentrated forces.

References

[2-1] A.H. Cottrell, Tewksbury Symposium on Fracture, University of Melbourne, 1963, p. 1.

[2-2] A.A. Griffith, The phenomena of rupture and flow in solids, Phil. Trans. Roy. Soc. (London) A221 (1920) 163–198.

[2-3] C.E. Inglis, Stresses in a plate due to the presence of cracks and sharp corners, Trans. Inst. Naval Architects 55 (1913) 219–230.

[2-4] E. Orowan, Notch brittleness and strength of metals, Trans. Inst. Engrs. Shipbuilders Scot. 89 (1945) 165–215.

[2-5] G.R. Irwin, Fracture dynamics, in: Fracture of Metals, ASM, Cleveland, OH, 1948, pp. 147–166.

PROBLEMS

2.1 Consider the cracked beam subjected to uniaxial tension shown in Figures 2.8 and 2.9. Find the strain energy release rate (per crack tip). Consider both fixed-end and constant force boundary conditions.

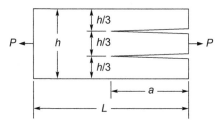

FIGURE 2.8

A cracked beam subjected to tension.

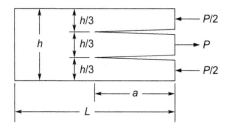

FIGURE 2.9

A cracked beam subjected to tension and compression.

2.2 Find the strain energy release rate G for the cracked beam shown in Figures 2.10 and 2.11. Use simple beam theory to model the cracked and uncracked regions. The thickness of the beam is t.

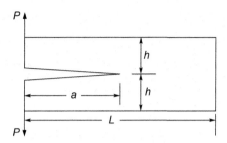

FIGURE 2.10

A cracked beam subjected to concentrated forces.

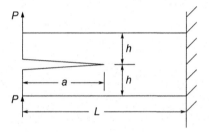

FIGURE 2.11

A cracked beam subjected to concentrated forces.

2.3 A cracked beam is subjected to a pair of forces at the center of the crack (see Figure 2.12). Find the minimum P that can split the beam. Assume $E = 70\,\text{GPa}$ and $G_c = 200\,\text{N}\cdot\text{m/m}^2$.

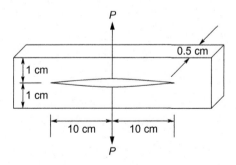

FIGURE 2.12

A center-cracked beam.

2.4 Find the strain energy release rate for the problem shown in Figure 2.13 wherein a thin elastic film of unit width is peeled from a rigid surface.

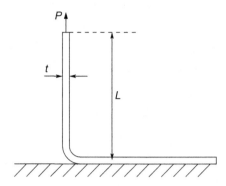

FIGURE 2.13

A thin film peeled from the rigid substrate.

2.5   Assume that the bending rigidity of the film is negligible, that L is large, and that the elastic constants of the film are known. The thin film is pulled parallel to the rigid surface as shown in Figure 2.14. Compare the strain energy release rate and the strain energy gained by the film during crack extension for both problems in Figures 2.13 and 2.14. For the problem of Figure 2.13, why is the strain energy released not the same as the strain energy gained by the film?

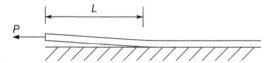

FIGURE 2.14

A thin film pulled parallel to the surface of the rigid substrate.

2.6   Show that the area $\overline{OBC}$ in Figure 2.6 does not depend on the loading condition during crack extension.

The Elastic Stress Field around a Crack Tip

Brittle fracture in a solid in the form of crack growth is governed by the stress field around the crack tip and by parameters that describe the resistance of the material to crack growth. Thus, the analysis of stresses near the crack tip constitutes an essential part of fracture mechanics. For brittle materials exhibiting linear elastic behavior, methods of elasticity are used to obtain stresses and displacements in cracked bodies. These methods include analytical ones, such as the complex potential function method and the integral transform method, and numerical ones, such as the finite element method. In this chapter, the complex potential function method is introduced and used to analyze the stresses and displacements around crack tips. The characteristics of the near-tip asymptotic stress and displacement fields and the crack growth criterion based on the crack tip field are discussed.

3.1 BASIC MODES OF FRACTURE AND STRESS INTENSITY FACTOR

A crack in a solid consists of disjoined upper and lower faces. The joint of the two crack faces forms the crack front. The two crack faces are usually assumed to lie in the same surface before deformation. When the cracked body is subjected to external loads (remotely or at the crack surfaces), the two crack faces move with respect to each other and these movements may be described by the differences in displacements u_x, u_y, and u_z between the upper and lower crack surfaces, where (x, y, z) is a local Cartesian coordinate system centered at the crack front with the x-axis perpendicular to the crack front, the y-axis perpendicular to the crack plane, and the z-axis along the crack front.

There are three independent movements corresponding to three fundamental fracture modes as pointed out by Irwin [3-1], which are schematically illustrated in Figure 3.1. These basic fracture modes are usually called Mode I, Mode II, and Mode III, respectively, and any fracture mode in a cracked body may be described by one of the three basic modes, or their combinations (see their descriptions on the next page).

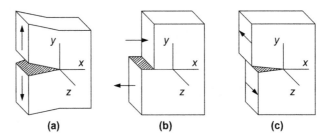

FIGURE 3.1

Schematic of the basic fracture modes: (a) Mode I (opening), (b) Mode II (sliding), (c) Mode III (tearing).

1. *Mode I (Opening Mode):* The two crack surfaces experience a jump only in u_y, that is, they move away symmetrically with respect to the undeformed crack plane (*xz*-plane).

2. *Mode II (Sliding Mode):* The two crack surfaces experience a jump only in u_x, that is, they slide against each other along directions perpendicular to the crack front but in the same undeformed plane.

3. *Mode III (Tearing Mode):* The two crack surfaces experience a jump only in u_z, that is, they tear over each other in the directions parallel to the crack front but in the same undeformed plane.

The three basic modes of crack deformation can be more precisely defined by the associated stresses ahead of the crack front, which may be considered as the crack tip in two-dimensional problems. It will be seen in the following sections that the near-tip stresses in the crack plane (*xz*-plane) for these three modes can be expressed as ($y = 0$, $x \rightarrow 0^+$)

$$\sigma_{yy} = \frac{K_I}{\sqrt{2\pi x}} + O(\sqrt{x}), \quad \sigma_{xy} = \sigma_{yz} = 0$$

$$\sigma_{xy} = \frac{K_{II}}{\sqrt{2\pi x}} + O(\sqrt{x}), \quad \sigma_{yy} = \sigma_{yz} = 0 \qquad (3.1)$$

$$\sigma_{yz} = \frac{K_{III}}{\sqrt{2\pi x}} + O(\sqrt{x}), \quad \sigma_{yy} = \sigma_{xy} = 0$$

respectively, where the three parameters K_I, K_{II}, and K_{III} are named stress intensity factors corresponding to the opening, sliding, and tearing (anti-plane shearing) modes of fracture, respectively.

These expressions show that the stresses have an inverse square root singularity at the crack tip and the stress intensity factors K_I, K_{II}, and K_{III} measure the intensities of the singular stress fields of opening, in-plane shearing, and anti-plane shearing, respectively. The stress intensity factor is a new concept in mechanics of solids and

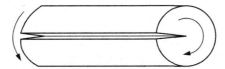

FIGURE 3.2

Mode III deformation in a cracked cylinder under torsion.

plays an essential role in the study of fracture strength of cracked solids. Various methods for determining stress intensity factors, including analytical, numerical, and experimental approaches, have been developed in the past few decades.

It is important to note that, except for Mode I deformation shown in Figure 3.1(a), the loading and specimen geometries shown in Figures 3.1(b) and (c) cannot be used to produce pure Mode II and Mode III deformation, respectively. In fact, unless an additional loading or boundary condition is specified, the cracked bodies cannot be in equilibrium. Other types of specimen are usually used. For instance, a long cylinder of a circular cross-section with a longitudinal slit under torsion (see Figure 3.2) can be used to produce a pure Mode III crack deformation.

3.2 METHOD OF COMPLEX POTENTIAL FOR PLANE ELASTICITY (THE KOLOSOV-MUSKHELISHVILI FORMULAS)

Among various mathematical methods in plane elasticity, the complex potential function method by Kolosov and Muskhelishvili [3-2] are one of the powerful and convenient methods to treat two-dimensional crack problems. In the complex potential method, stresses and displacements are expressed in terms of analytic functions of complex variables. The problem of obtaining stresses and displacements around a crack tip is converted to finding some analytic functions subjected to appropriate boundary conditions. A brief introduction of the general formulation of the Kolosov and Muskhelishvili complex potentials is given in this section.

3.2.1 Basic Equations of Plane Elasticity and Airy Stress Function

The basic equations of elasticity consist of equilibrium equations of stresses, strain-displacement relations, and Hooke's law that relates stresses and strains. In plane elasticity (plane strain and plane stress), the equilibrium equations are (body forces are absent)

$$\frac{\partial \sigma_{xx}}{\partial x} + \frac{\partial \sigma_{xy}}{\partial y} = 0$$

$$\frac{\partial \sigma_{xy}}{\partial x} + \frac{\partial \sigma_{yy}}{\partial y} = 0$$

$$(3.2)$$

where σ_{xx}, σ_{yy}, and σ_{xy} are stresses, and (x,y) are Cartesian coordinates. The strains and the displacements are related by

$$e_{xx} = \frac{\partial u_x}{\partial x}, \quad e_{yy} = \frac{\partial u_y}{\partial y}, \quad e_{xy} = \frac{1}{2}\left(\frac{\partial u_x}{\partial y} + \frac{\partial u_y}{\partial x}\right) \tag{3.3}$$

where e_{xx}, e_{yy}, and e_{xy} are tensorial strain components, and u_x and u_y are displacements. The stress–strain relations are given by

$$\sigma_{xx} = \lambda^* \left(e_{xx} + e_{yy}\right) + 2\mu e_{xx}$$
$$\sigma_{yy} = \lambda^* \left(e_{xx} + e_{yy}\right) + 2\mu e_{yy} \tag{3.4}$$
$$\sigma_{xy} = 2\mu e_{xy}$$

or inversely

$$e_{xx} = \frac{1}{2\mu}\left[\sigma_{xx} - \frac{\lambda^*}{2\left(\lambda^* + \mu\right)}\left(\sigma_{xx} + \sigma_{yy}\right)\right]$$
$$e_{yy} = \frac{1}{2\mu}\left[\sigma_{yy} - \frac{\lambda^*}{2\left(\lambda^* + \mu\right)}\left(\sigma_{xx} + \sigma_{yy}\right)\right] \tag{3.5}$$
$$e_{xy} = \frac{1}{2\mu}\sigma_{xy}$$

where μ is the shear modulus and

$$\lambda^* = \frac{3 - \kappa}{\kappa - 1}\mu$$

in which

$$\kappa = \begin{cases} 3 - 4\nu & \text{for plane strain} \\ \dfrac{3 - \nu}{1 + \nu} & \text{for plane stress} \end{cases} \tag{3.6}$$

In the previous relation, ν is Poisson's ratio. The compatibility equation of strains can be obtained from Eq. (3.3) by eliminating the displacements as follows:

$$\frac{\partial^2 e_{xx}}{\partial y^2} + \frac{\partial^2 e_{yy}}{\partial x^2} = 2\frac{\partial^2 e_{xy}}{\partial x \partial y} \tag{3.7}$$

By using the stress–strain relations Eq. (3.5) together with the equations of equilibrium Eq. (3.2), the compatibility condition Eq. (3.7) can be expressed in terms of stresses as

$$\nabla^2 (\sigma_{xx} + \sigma_{yy}) = 0 \tag{3.8}$$

where

$$\nabla^2 = \frac{\partial^2}{\partial x^2} + \frac{\partial^2}{\partial y^2}$$

is the Laplace operator.

The Airy stress function ϕ is defined through

$$\sigma_{xx} = \frac{\partial^2 \phi}{\partial y^2}, \quad \sigma_{xy} = -\frac{\partial^2 \phi}{\partial x \partial y}, \quad \sigma_{yy} = \frac{\partial^2 \phi}{\partial x^2} \tag{3.9}$$

Using these relations, the equilibrium equations in Eq. (3.2) are automatically satisfied, and the compatibility Eq. (3.8) becomes

$$\nabla^4 \phi = \nabla^2 \nabla^2 \phi = 0 \tag{3.10}$$

where

$$\nabla^4 = \nabla^2 \nabla^2 = \frac{\partial^4}{\partial x^4} + 2\frac{\partial^4}{\partial x^2 \partial y^2} + \frac{\partial^4}{\partial y^4}$$

is the biharmonic operator. Any function ϕ satisfying Eq. (3.10) is called a biharmonic function. A harmonic function f satisfies $\nabla^2 f = 0$. Thus, if f is harmonic, it is also biharmonic. However, the converse is not true. Once the Airy stress function is known, the stresses can be obtained by Eq. (3.9) and strains and displacements obtained through Eqs. (3.5) and (3.3), respectively.

3.2.2 Analytic Functions and Cauchy-Riemann Equations

In a Cartesian coordinate system (x, y), the complex variable z and its conjugate $\bar{z}$ are defined as

$$z = x + iy$$

and

$$\bar{z} = x - iy$$

respectively, where $i = \sqrt{-1}$. They can also be expressed in polar coordinates (r, θ) as

$$z = r(\cos\theta + i\sin\theta) = re^{i\theta}$$

and

$$\bar{z} = r(\cos\theta - i\sin\theta) = re^{-i\theta}$$

respectively.

Consider a function of the complex variable z, $f(z)$. The derivative of $f(z)$ with respect to z is by definition

$$\frac{df(z)}{dz} = \lim_{\Delta z \to 0} \frac{f(z+\Delta z) - f(z)}{\Delta z}$$

If $f(z)$ has a derivative at point z_0 and also at each point in some neighborhood of z_0, then $f(z)$ is said to be analytic at z_0. The complex function $f(z)$ can be expressed in the form

$$f(z) = u(x,y) + iv(x,y)$$

where u and v are real functions. If $f(z)$ is analytic, we have

$$\frac{\partial}{\partial x} f(z) = f'(z)\frac{\partial z}{\partial x} = f'(z)$$

and

$$\frac{\partial}{\partial y} f(z) = f'(z)\frac{\partial z}{\partial y} = if'(z)$$

where a prime stands for differentiation with respect to z. Thus,

$$\frac{\partial}{\partial x} f(z) = -i\frac{\partial}{\partial y} f(z)$$

or

$$\frac{\partial u}{\partial x} + i\frac{\partial v}{\partial x} = \frac{\partial v}{\partial y} - i\frac{\partial u}{\partial y}$$

From this equation, we obtain the Cauchy-Riemann equations:

$$\frac{\partial u}{\partial x} = \frac{\partial v}{\partial y}, \quad \frac{\partial u}{\partial y} = -\frac{\partial v}{\partial x} \tag{3.11}$$

These equations can also be shown to be sufficient for $f(z)$ to be analytic.

From the Cauchy-Riemann equations it is easy to derive the following:

$$\nabla^2 u = \nabla^2 v = 0$$

that is, the real and imaginary parts of an analytic function are harmonic.

3.2.3 Complex Potential Representation of the Airy Stress Function

The Airy stress function ϕ is biharmonic according to Eq. (3.10). Introduce a function P by

$$\nabla^2 \phi = P \tag{3.12}$$

then

$$\nabla^2 P = \nabla^2 \nabla^2 \phi = 0$$

This simply says that P is a harmonic function. Hence,

$$P = \text{Real part of } f(z) \equiv \text{Re}\{f(z)\}$$

where $f(z)$ is an analytic function and can be expressed as

$$f(z) = P + iQ$$

Let

$$\psi(z) = \frac{1}{4} \int f(z) dz = p + iq$$

then ψ is also analytic and its derivative is given by

$$\psi'(z) = \frac{1}{4} f(z)$$

According to the Cauchy-Riemann equations, we have

$$\psi'(z) = \frac{\partial p}{\partial x} + i \frac{\partial q}{\partial x} = \frac{\partial q}{\partial y} - i \frac{\partial p}{\partial y}$$

A relation between P and p (or q) can then be obtained:

$$P = 4 \frac{\partial p}{\partial x} = 4 \frac{\partial q}{\partial y} \tag{3.13}$$

Consider the function $\phi - (xp + yq)$. It can be shown that

$$\nabla^2 [\phi - (xp + yq)] = 0$$

Thus, $\phi - (xp + yq)$ is harmonic and is a real (or imaginary) part of an analytic function, say $\chi(z)$, that is,

$$\phi - (xp + yq) = \text{Re}\{\chi(z)\}$$

Using the relation

$$xp + yq = \text{Re}\{\bar{z}\psi(z)\}$$

we obtain the complex potential representation of the Airy stress function

$$\phi = \text{Re}\{\bar{z}\psi(z) + \chi(z)\}$$
$$2\phi(x, y) = \bar{z}\psi(z) + z\overline{\psi(z)} + \chi(z) + \overline{\chi(z)} \tag{3.14}$$

3.2.4 Stress and Displacement

From the definition of the Airy stress function we obtain

$$\sigma_{xx} + i\sigma_{xy} = \frac{\partial^2 \phi}{\partial y^2} - i\frac{\partial^2 \phi}{\partial x \partial y} = -i\frac{\partial}{\partial y}\left(\frac{\partial \phi}{\partial x} + i\frac{\partial \phi}{\partial y}\right)$$

$$\sigma_{yy} - i\sigma_{xy} = \frac{\partial^2 \phi}{\partial x^2} + i\frac{\partial^2 \phi}{\partial x \partial y} = \frac{\partial}{\partial x}\left(\frac{\partial \phi}{\partial x} + i\frac{\partial \phi}{\partial y}\right)$$

(3.15)

Note that for an analytic function $f(z)$ we have

$$\frac{\partial f(z)}{\partial x} = f'(z)\frac{\partial z}{\partial x} = f'(z)$$

$$\frac{\overline{\partial f(z)}}{\partial x} = \overline{\left(\frac{\partial f(z)}{\partial x}\right)} = \overline{f'(z)}$$

$$\frac{\partial f(z)}{\partial y} = f'(z)\frac{\partial z}{\partial y} = if'(z)$$

$$\frac{\overline{\partial f(z)}}{\partial y} = \overline{\left(\frac{\partial f(z)}{\partial y}\right)} = -i\overline{f'(z)}$$

Using the preceding relations together with Eq. (3.14), we obtain

$$\frac{\partial \phi}{\partial x} + i\frac{\partial \phi}{\partial y} = \psi(z) + z\overline{\psi'(z)} + \overline{\chi'(z)}$$

(3.16)

Substitution of the relation in (3.16) in Eq. (3.15) leads to

$$\sigma_{xx} + i\sigma_{xy} = \psi'(z) + \overline{\psi'(z)} - z\overline{\psi''(z)} - \overline{\chi''(z)}$$

$$\sigma_{yy} - i\sigma_{xy} = \psi'(z) + \overline{\psi'(z)} + z\overline{\psi''(z)} + \overline{\chi''(z)}$$

(3.17)

Summing the two equations in Eq. (3.17), we have

$$\sigma_{xx} + \sigma_{yy} = 2[\psi'(z) + \overline{\psi'(z)}] = 4\,\mathrm{Re}[\psi'(z)]$$

(3.18)

Subtracting the first equation from the second one in Eq. (3.17), we obtain

$$\sigma_{yy} - \sigma_{xx} - 2i\sigma_{xy} = 2z\overline{\psi''(z)} + 2\overline{\chi''(z)}$$

The equation above can be rewritten in the following form by taking the conjugate of the quantities on both sides:

$$\sigma_{yy} - \sigma_{xx} + 2i\sigma_{xy} = 2[\bar{z}\psi''(z) + \chi''(z)]$$

(3.19)

Equations (3.18) and (3.19) are the convenient analytic function representations of stresses.

We now turn to the complex potential representation of displacements. Substituting the strain-displacement relations Eq. (3.3) and the stresses in Eq. (3.9) into the stress–strain relations Eq. (3.5) yields

$$2\mu \frac{\partial u_x}{\partial x} = \frac{\partial^2 \phi}{\partial y^2} - \frac{\lambda^*}{2(\lambda^* + \mu)} \nabla^2 \phi$$

$$2\mu \frac{\partial u_y}{\partial y} = \frac{\partial^2 \phi}{\partial x^2} - \frac{\lambda^*}{2(\lambda^* + \mu)} \nabla^2 \phi \qquad (3.20)$$

$$\mu \left(\frac{\partial u_x}{\partial y} + \frac{\partial u_y}{\partial x} \right) = -\frac{\partial^2 \phi}{\partial x \partial y}$$

From Eqs. (3.12) and (3.13) we have

$$\nabla^2 \phi = P = 4 \frac{\partial p}{\partial x} = 4 \frac{\partial q}{\partial y}$$

Substitution of this equation into the first two equations in Eq. (3.20) yields

$$2\mu \frac{\partial u_x}{\partial x} = -\frac{\partial^2 \phi}{\partial x^2} + \frac{2(\lambda^* + 2\mu)}{\lambda^* + \mu} \frac{\partial p}{\partial x}$$

$$2\mu \frac{\partial u_y}{\partial y} = -\frac{\partial^2 \phi}{\partial y^2} + \frac{2(\lambda^* + 2\mu)}{\lambda^* + \mu} \frac{\partial q}{\partial y}$$

Integrating the preceding equations, we obtain

$$2\mu u_x = -\frac{\partial \phi}{\partial x} + \frac{2(\lambda^* + 2\mu)}{\lambda^* + \mu} p + f_1(y)$$

$$2\mu u_y = -\frac{\partial \phi}{\partial y} + \frac{2(\lambda^* + 2\mu)}{\lambda^* + \mu} q + f_2(x) \qquad (3.21)$$

Substituting these expressions in the third equation in Eq. (3.20), we can conclude that $f_1(y)$ and $f_2(x)$ represent rigid body displacements and thus can be neglected. Rewrite Eq. (3.21) in complex form:

$$2\mu(u_x + i u_y) = -\left(\frac{\partial \phi}{\partial x} + i \frac{\partial \phi}{\partial y} \right) + \frac{2(\lambda^* + 2\mu)}{\lambda^* + \mu} \psi(z)$$

Using Eq. (3.16) in the previous expression, we arrive at the complex potential representation of displacements:

$$2\mu(u_x + i u_y) = \kappa \psi(z) - z \overline{\psi'(z)} - \overline{\chi'(z)} \qquad (3.22)$$

In deriving Eq. (3.22), the relation

$$\kappa = \frac{\lambda^* + 3\mu}{\lambda^* + \mu}$$

is used. Equations (3.18), (3.19), and (3.22) are the Kolosov-Muskhelishvili formulas.

3.3 WESTERGAARD FUNCTION METHOD

The Kolosov-Muskhelishvili formulas hold for general plane elasticity problems. In applications to crack problems, however, the three basic fracture modes discussed in Section 3.1 possess symmetry or antisymmetry properties. The Westergaard function method [3-3, 3-4] is more convenient for discussing these basic crack problems. We will introduce the Westergaard functions using the general Kolosov-Muskhelishvili formulas, which are rewritten here for convenience:

$$\sigma_{xx} + \sigma_{yy} = 4\,\mathrm{Re}\{\psi'(z)\} \tag{3.23}$$

$$\sigma_{yy} - \sigma_{xx} + 2i\sigma_{xy} = 2\{\bar{z}\psi''(z) + \chi''(z)\} \tag{3.24}$$

$$2\mu(u_x + iu_y) = \kappa\psi(z) - z\overline{\psi'(z)} - \overline{\chi'(z)} \tag{3.25}$$

3.3.1 Symmetric Problems (Mode I)

Consider an infinite plane with cracks along the x-axis. If the external loads are symmetric with respect to the x-axis, then $\sigma_{xy} = 0$ along $y = 0$. From Eq. (3.24), we have

$$\mathrm{Im}\{\bar{z}\psi''(z) + \chi''(z)\} = 0 \quad \text{at } y = 0 \tag{3.26}$$

The preceding equation can be satisfied if and only if

$$\chi''(z) + z\psi''(z) + A = 0 \tag{3.27}$$

in which A is a real constant.

Proof

It is clear that Eq. (3.27) leads to Eq. (3.26) because $z = \bar{z}$ at $y = 0$. For the converse case, consider

$$\chi''(z) + z\psi''(z) = -A(z) \tag{3.28}$$

We now prove from Eq. (3.26) that $A(z)$ is a real constant. First, $A(z)$ is analytic since χ'' and $z\psi''$ are analytic. Second, $A(z)$ is bounded in the entire plane according to Eq. (3.24) (stresses are finite everywhere except at crack tips) and the general asymptoptic solutions of $\psi(z)$ and $\chi(z)$ at crack tips (see Section 3.6). $A(z)$ is thus a constant $(= A)$ by Liouville's theorem. Substituting Eq. (3.28) into the left side of Eq. (3.26), we have

$$\mathrm{Im}\{(\bar{z} - z)\psi''(z) - A\} = \mathrm{Im}\{-2iy\psi''(z) - A\} = 0 \quad \text{at } y = 0$$

or

$$\mathrm{Im}\{-A\} = 0$$

Hence, we can conclude that

$$A = \text{real constant}$$

Because $\psi(z)$ and $\chi(z)$ are related according to Eq. (3.27), stresses and displacements may be expressed by only one of the two analytic functions. From Eq. (3.27), we obtain

$$\chi''(z) = -z\psi'' - A$$

Substituting this equation into Eqs. (3.23) through (3.25) and solving the resulting equations, we have

$$\sigma_{xx} = 2\,\mathrm{Re}\{\psi'\} - 2y\,\mathrm{Im}\{\psi''\} + A$$

$$\sigma_{yy} = 2\,\mathrm{Re}\{\psi'\} + 2y\,\mathrm{Im}\{\psi''\} - A$$

$$\sigma_{xy} = -2y\,\mathrm{Re}\{\psi''\} \qquad (3.29)$$

$$2\mu u_x = (\kappa - 1)\,\mathrm{Re}\{\psi\} - 2y\,\mathrm{Im}\{\psi'\} + Ax$$

$$2\mu u_y = (\kappa + 1)\,\mathrm{Im}\{\psi\} - 2y\,\mathrm{Re}\{\psi'\} - Ay$$

Define

$$\psi' = \frac{1}{2}(Z_I + A)$$

Thus,

$$\psi = \frac{1}{2}(\hat{Z}_I + Az)$$

$$\psi'' = \frac{1}{2}Z_I'$$

where $\hat{Z}_I' \equiv Z_I$. The use of these two equations in Eq. (3.29) results in

$$\sigma_{xx} = \mathrm{Re}\{Z_I\} - y\,\mathrm{Im}\{Z_I'\} + 2A$$

$$\sigma_{yy} = \mathrm{Re}\{Z_I\} + y\,\mathrm{Im}\{Z_I'\}$$

$$\sigma_{xy} = -y\,\mathrm{Re}\{Z_I'\} \qquad (3.30)$$

$$2\mu u_x = \frac{(\kappa - 1)}{2}\,\mathrm{Re}\{\hat{Z}_I\} - y\,\mathrm{Im}\{Z_I\} + \frac{1}{2}(\kappa + 1)Ax$$

$$2\mu u_y = \frac{(\kappa + 1)}{2}\,\mathrm{Im}\{\hat{Z}_I\} - y\,\mathrm{Re}\{Z_I\} + \frac{1}{2}(\kappa - 3)Ay$$

Z_I is the so-called Westergaard function for Mode I problems. It is obvious that the stress field associated with A is a uniform uniaxial stress $\sigma_{xx} = 2A$. This stress field does not add to the stress singularity at the crack tip.

From Eq. (3.30), we note that

$$\sigma_{xx} - \sigma_{yy} = 2A = \text{constant} \quad \text{at } y = 0$$

For the case where uniform tension is applied in the y-direction; that is, $\sigma_{xx} = 0$ and $\sigma_{yy} = \sigma_0, A = -\sigma_0/2$. If the panel is subjected to biaxial tensions of equal magnitude, then $A = 0$.

The Airy stress function corresponding to the constant stress field $\sigma_{xx} = 2A = \partial^2\phi/\partial y^2$ is given by

$$\phi = Ay^2$$

It can be easily verified that the stresses and displacements of Eq. (3.30) are derived from the stress function (see Problem 3.3)

$$\phi = \text{Re}\{\widehat{\widehat{Z}}_I\} + y\,\text{Im}\{\widehat{Z}_I\} + Ay^2$$

The function Z_I is usually associated with Westergaard [3-3], who used it to solve the contact pressure distribution resulting from the contact of many surfaces and some crack problems. In its original form used by Westergaard, $A = 0$. In 1957, Irwin [3-1] used Westergaard's solutions to obtain the stress field at the crack tip and related that to the strain energy release rate.

3.3.2 Skew-Symmetric Problems (Mode II)

For loads that are skew-symmetric with respect to the crack line (x-axis), the normal stress σ_{yy} is zero along $y = 0$. From Eqs. (3.23) and (3.24), this condition gives rise to

$$\text{Re}\{2\psi'(z) + \bar{z}\psi''(z) + \chi''\} = 0 \quad \text{at } y = 0 \tag{3.31}$$

Following the same procedure described for the symmetric problem, we obtain

$$\chi''(z) + 2\psi'(z) + z\psi''(z) + iB = 0$$

in which B is a real constant. Using this equation, $\chi(z)$ can be eliminated in Eqs. (3.23) through (3.25) and we have

$$\begin{aligned}
\sigma_{xx} &= 4\,\text{Re}\{\psi'(z)\} - 2y\,\text{Im}\{\psi''(z)\} \\
\sigma_{yy} &= 2y\,\text{Im}\{\psi''(z)\} \\
\sigma_{xy} &= -2\,\text{Im}\{\psi'(z)\} - 2y\,\text{Re}\{\psi''(z)\} - B \\
2\mu u_x &= (\kappa+1)\,\text{Re}\{\psi(z)\} - 2y\,\text{Im}\{\psi'(z)\} - By \\
2\mu u_y &= (\kappa-1)\text{Im}\{\psi(z)\} - 2y\,\text{Re}\{\psi'(z)\} - Bx
\end{aligned} \tag{3.32}$$

Define an analytic function $\Psi(z)$ by

$$\Psi'(z) = \psi'(z) + \frac{i}{2}B$$

Then

$$\Psi(z) = \psi(z) + \frac{i}{2}Bz$$

$$\Psi''(z) = \psi''(z)$$

Substituting these definitions into Eq. (3.32), we obtain

$$\sigma_{xx} = 4\,\mathrm{Re}\{\Psi'(z)\} - 2y\mathrm{Im}\{\Psi''(z)\}$$
$$\sigma_{yy} = 2y\,\mathrm{Im}\{\Psi''\}$$
$$\sigma_{xy} = -2\,\mathrm{Im}\{\Psi'(z)\} - 2y\,\mathrm{Re}\{\Psi''(z)\} \tag{3.33}$$
$$2\mu u_x = (\kappa + 1)\mathrm{Re}\{\Psi(z)\} - 2y\,\mathrm{Im}\{\Psi'(z)\} + \frac{\kappa + 1}{2}By$$
$$2\mu u_x = (\kappa - 1)\mathrm{Im}\{\Psi(z)\} - 2y\,\mathrm{Re}\{\Psi'(z)\} - \frac{\kappa + 1}{2}Bx$$

The last term in the displacement components u_x and u_y represents a rigid body rotation. Define Westergaard function Z_{II} as

$$Z_{II} = 2i\Psi'(z)$$

Then Eq. (3.33) becomes

$$\sigma_{xx} = 2\mathrm{Im}\{Z_{II}\} + y\,\mathrm{Re}\{Z_{II}'\}$$
$$\sigma_{yy} = -y\,\mathrm{Re}\{Z_{II}'\}$$
$$\sigma_{xy} = \mathrm{Re}\{Z_{II}\} - y\mathrm{Im}\{Z_{II}'\} \tag{3.34}$$
$$2\mu u_x = \frac{1}{2}(\kappa + 1)\mathrm{Im}\{\widehat{Z}_{II}\} + y\,\mathrm{Re}\{Z_{II}\} + \frac{\kappa + 1}{2}By$$
$$2\mu u_y = -\frac{1}{2}(\kappa - 1)\mathrm{Re}\{\widehat{Z}_{II}\} - y\,\mathrm{Im}\{Z_{II}\} - \frac{\kappa + 1}{2}Bx$$

Thus, the Westergaard function $Z_{II}(z)$ provides the general solution for the skew-symmetric problems.

We conclude that any plane elasticity problem involving collinear straight cracks in an infinite plane can be completely solved by the stress function

$$\phi = \mathrm{Re}\{\widehat{\widehat{Z}}_I\} + y\mathrm{Im}\{\widehat{Z}_I\} - y\,\mathrm{Re}\{\widehat{Z}_{II}\} + Ay^2$$

This is an alternative form to Eq. (3.14). The advantage of using the Westergaard functions is that the two modes of fracture are represented separately by two analytic functions.

3.4 SOLUTIONS BY THE WESTERGAARD FUNCTION METHOD

A center crack in an infinite plate under uniform remote loading is perhaps the best example to introduce the basic concepts of stress intensity factor and near-tip stress and deformation fields. In this section, the Westergaard function method is used to find the elasticity solutions for an infinite plane with a center crack under uniform biaxial tension, in-plane shear, and antiplane shear loading, respectively. We will see that the near-tip singular stress fields can be easily extracted from the complete solutions and the stress intensity factors can be obtained from the solutions using the definition in Eq. (3.1). Furthermore, stress intensity factors may also be conveniently determined directly from the general Kolosov-Muskhelishvili potentials or Westergaard functions.

3.4.1 Mode I Crack

One of the most typical crack problems in fracture mechanics is an infinite plane with a line crack of length $2a$ subjected to biaxial stress σ_0 at infinity, as shown in Figure 3.3. In practice, if the crack length is much smaller than any in-plane size of the concerned elastic body, the region may be mathematically treated as an infinite plane with a finite crack. The problem is Mode I since the loads are symmetric with respect to the crack line.

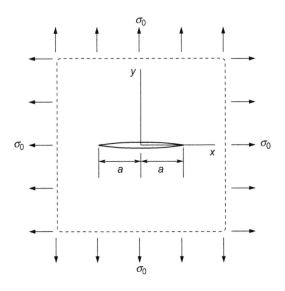

FIGURE 3.3

A crack in an infinite elastic plane subjected to biaxial tension.

Solution of Stresses

The boundary conditions of the crack problem are

$$\sigma_{xy} = \sigma_{yy} = 0 \quad \text{at } |x| \le a \text{ and } y = 0 \quad \text{(crack surfaces)}$$

$$\sigma_{xx} = \sigma_{yy} = \sigma_0, \ \sigma_{xy} = 0 \quad \text{at } x^2 + y^2 \to \infty \tag{3.35}$$

The solution to the preceding boundary value problem was given by Westergaard [3-3] with

$$Z_I(z) = \frac{\sigma_0 z}{\sqrt{z^2 - a^2}}, \quad A = 0 \tag{3.36}$$

Since all the equations of plane elasticity are automatically satisfied, we only need to verify that the stresses obtained by Eq. (3.30) with Z_I given in Eq. (3.36) satisfy the boundary conditions Eq. (3.35).

To find the explicit expressions of the stress field, it is more convenient to use the polar coordinates shown in Figure 3.4. The following relations are obvious:

$$z = r e^{i\theta}$$

$$z - a = r_1 e^{i\theta_1} \tag{3.37}$$

$$z + a = r_2 e^{i\theta_2}$$

In terms of these polar coordinates, the function Z_I becomes

$$Z_I = \frac{\sigma_0 z}{\sqrt{z+a}\sqrt{z-a}}$$

$$= \frac{\sigma_0 r}{\sqrt{r_1 r_2}} \exp i\left(\theta - \frac{1}{2}\theta_1 - \frac{1}{2}\theta_2\right) \tag{3.38}$$

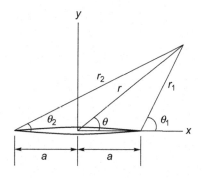

FIGURE 3.4

Polar coordinate systems.

The derivative of Z_I is obtained as

$$Z_I' = \frac{\sigma_0}{\sqrt{z^2 - a^2}} - \frac{\sigma_0 z^2}{(z^2 - a^2)^{3/2}} = -\frac{\sigma_0 a^2}{(z^2 - a^2)^{3/2}}$$

$$= -\frac{\sigma_0 a^2}{(r_1 r_2)^{3/2}} \exp\left(-i\frac{3}{2}(\theta_1 + \theta_2)\right) \tag{3.39}$$

It follows from Eqs. (3.38) and (3.39) that

$$\text{Re}\{Z_I\} = \frac{\sigma_0 r}{\sqrt{r_1 r_2}} \cos\left(\theta - \frac{1}{2}\theta_1 - \frac{1}{2}\theta_2\right)$$

$$\text{Re}\{Z_I'\} = \frac{-\sigma_0 a^2}{(r_1 r_2)^{3/2}} \cos\frac{3}{2}(\theta_1 + \theta_2)$$

$$\text{Im}\{Z_I'\} = \frac{\sigma_0 a^2}{(r_1 r_2)^{3/2}} \sin\frac{3}{2}(\theta_1 + \theta_2)$$

Using these expressions, the normal stress σ_{xx} is obtained as

$$\sigma_{xx} = \text{Re}\{Z_I\} - y\,\text{Im}\{Z_I'\}$$

$$= \frac{\sigma_0 r}{\sqrt{r_1 r_2}} \cos\left(\theta - \frac{1}{2}\theta_1 - \frac{1}{2}\theta_2\right) - \frac{\sigma_0 a^2}{(r_1 r_2)^{3/2}} r\sin\theta \sin\frac{3}{2}(\theta_1 + \theta_2)$$

$$= \frac{\sigma_0 r}{\sqrt{r_1 r_2}} \left[\cos\left(\theta - \frac{1}{2}\theta_1 - \frac{1}{2}\theta_2\right) - \frac{a^2}{r_1 r_2} \sin\theta \sin\frac{3}{2}(\theta_1 + \theta_2)\right] \tag{3.40}$$

Similarly, the other two stress components can be obtained:

$$\sigma_{yy} = \frac{\sigma_0 r}{\sqrt{r_1 r_2}} \left[\cos\left(\theta - \frac{1}{2}\theta_1 - \frac{1}{2}\theta_2\right) + \frac{a^2}{r_1 r_2} \sin\theta \sin\frac{3}{2}(\theta_1 + \theta_2)\right] \tag{3.41}$$

$$\sigma_{xy} = \frac{\sigma_0 r}{\sqrt{r_1 r_2}} \left[\frac{a^2}{r_1 r_2} \sin\theta \cos\frac{3}{2}(\theta_1 + \theta_2)\right] \tag{3.42}$$

Using these stress expressions, it is easy to show that the tractions (σ_{yy} and σ_{xy}) on the crack surface ($\theta_1 = \pi, \theta_2 = 0, \theta = 0, \pi$) vanish completely.

At large distances ($r \to \infty$) from the crack it is easily seen that

$$r_1 \approx r_2 \approx r \to \infty$$

$$\theta_1 \approx \theta_2 \approx \theta$$

and consequently that

$$\sigma_{xx} = \sigma_0, \quad \sigma_{yy} = \sigma_0, \quad \sigma_{xy} = 0$$

Thus, the boundary conditions are satisfied everywhere. Equations (3.40) through (3.42) are the complete solution of the stress field in the entire cracked plane.

The Near-Tip Solution

In fracture mechanics, crack growth is controlled by the stresses and deformations around the crack tip. We thus study the near-tip asymptotic stress field. In the vicinity of the crack tip (say the right tip), we have

$$\frac{r_1}{a} \ll 1, \quad \theta \approx 0, \quad \theta_2 \approx 0$$

$$r \approx a, \quad r_2 \approx 2a$$

$$\sin\theta \approx \frac{r_1}{a}\sin\theta_1$$

$$\sin\frac{3}{2}(\theta_1 + \theta_2) \approx \sin\frac{3}{2}\theta_1$$

$$\cos\left(\theta - \frac{1}{2}\theta_1 - \frac{1}{2}\theta_2\right) \approx \cos\frac{1}{2}\theta_1$$

$$\cos\frac{3}{2}(\theta_1 + \theta_2) \approx \cos\frac{3}{2}\theta_1$$

By using the preceding asymptotic expressions, Eq. (3.40) reduces to

$$\begin{aligned}
\sigma_{xx} &= \frac{\sigma_0 a}{\sqrt{2ar_1}}\left(\cos\frac{1}{2}\theta_1 - \frac{a^2}{2ar_1}\frac{r_1}{a}\sin\theta_1\sin\frac{3}{2}\theta_1\right) \\
&= \frac{\sigma_0\sqrt{a}}{\sqrt{2r_1}}\left(\cos\frac{1}{2}\theta_1 - \frac{1}{2}\sin\theta_1\sin\frac{3}{2}\theta_1\right) \\
&= \frac{\sigma_0\sqrt{a}}{\sqrt{2r_1}}\cos\frac{1}{2}\theta_1\left(1 - \sin\frac{1}{2}\theta_1\sin\frac{3}{2}\theta_1\right)
\end{aligned}$$

Similarly,

$$\sigma_{yy} = \frac{\sigma_0\sqrt{a}}{\sqrt{2r_1}}\cos\frac{1}{2}\theta_1\left(1 + \sin\frac{1}{2}\theta_1\sin\frac{3}{2}\theta_1\right)$$

$$\sigma_{xy} = \frac{\sigma_0\sqrt{a}}{\sqrt{2r_1}}\sin\frac{1}{2}\theta_1\cos\frac{1}{2}\theta_1\cos\frac{3}{2}\theta_1$$

Along the crack extended line ($\theta = \theta_1 = \theta_2 = 0$), these near-tip stresses are

$$\sigma_{yy} = \frac{\sigma_0\sqrt{\pi a}}{\sqrt{2\pi r_1}}$$

$$\sigma_{xy} = 0$$

Comparing the previous stresses with Eq. (3.1) in Section 3.1, we have the Mode I stress intensity factor K_I for the crack problem as

$$K_I = \sigma_0 \sqrt{\pi a} \tag{3.43}$$

and the Mode II stress intensity factor $K_{II} = 0$.

If the origin of the coordinate system (r, θ) is located at the crack tip, then the stress field near the crack tip can be written in terms of the stress intensity factor K_I as

$$
\begin{aligned}
\sigma_{xx} &= \frac{K_I}{\sqrt{2\pi r}} \cos \frac{1}{2}\theta \left(1 - \sin \frac{1}{2}\theta \sin \frac{3}{2}\theta \right) \\
\sigma_{yy} &= \frac{K_I}{\sqrt{2\pi r}} \cos \frac{1}{2}\theta \left(1 + \sin \frac{1}{2}\theta \sin \frac{3}{2}\theta \right) \\
\sigma_{xy} &= \frac{K_I}{\sqrt{2\pi r}} \sin \frac{1}{2}\theta \cos \frac{1}{2}\theta \cos \frac{3}{2}\theta
\end{aligned}
\tag{3.44}
$$

Following the same procedures, the near-tip displacements are obtained as

$$
\begin{aligned}
u_x &= \frac{K_I}{8\mu\pi} \sqrt{2\pi r} \left[(2\kappa - 1)\cos \frac{\theta}{2} - \cos \frac{3\theta}{2} \right] \\
u_y &= \frac{K_I}{8\mu\pi} \sqrt{2\pi r} \left[(2\kappa + 1)\sin \frac{\theta}{2} - \sin \frac{3\theta}{2} \right]
\end{aligned}
\tag{3.45}
$$

Equation (3.44) shows that stresses have an inverse square root singularity at the crack tip and the intensity of the singular stress field is described by the stress intensity factor.

Equations (3.44) and (3.45) are derived for a crack in an infinite plane subjected to remote biaxial tension. It will be seen in Section 3.7 that these expressions for the near-tip stresses and displacements hold for any cracked body undergoing Mode I deformations. The difference is only the value of the stress intensity factor. Hence, once the stress σ_{yy} along the crack extended line is known, the Mode I stress intensity factor can be obtained from

$$K_I = \lim_{r \to 0} \sqrt{2\pi r} \, \sigma_{yy}(\theta = 0) \tag{3.46}$$

Crack Surface Displacement

Besides the near-tip stress field, the displacements of crack faces are also relevent to crack growth. Under Mode I deformation conditions, the crack surfaces open up, which is quantified by the vertical displacement component u_y. For the problem shown in Figure 3.3, we already have

$$Z_I = \frac{\sigma_0 z}{\sqrt{z^2 - a^2}}$$

and

$$\hat{Z}_I = \sigma_0 \sqrt{z^2 - a^2}$$

In terms of the polar coordinates defined in Eq. (3.37) and Figure 3.4, the function $\hat{Z}_I$ can be expressed as

$$\hat{Z}_I = \sigma_0 \sqrt{r_1 r_2} e^{\frac{1}{2} i(\theta_1 + \theta_2)}$$

The corresponding vertical displacement is obtained from the last equation of Eq. (3.30):

$$4\mu u_y = (\kappa + 1) \mathrm{Im} \hat{Z}_I$$

$$= (\kappa + 1) \sigma_0 \sqrt{r_1 r_2} \sin \frac{1}{2} (\theta_1 + \theta_2)$$

The upper crack surface corresponds to $\theta_1 = \pi$, $\theta_2 = 0$ and the lower surface is $\theta_1 = -\pi$, $\theta_2 = 0$. The displacement of the upper crack surface is thus given by

$$u_y = \frac{\kappa + 1}{4\mu} \sigma_0 \sqrt{r_1 r_2} = \frac{\kappa + 1}{4\mu} \sigma_0 \sqrt{a^2 - x^2} \tag{3.47}$$

Near the crack tip, $r_1 \ll a$, $r_2 \approx 2a$, and Eq. (3.47) reduces to

$$u_y = \frac{\kappa + 1}{4\mu} \sigma_0 \sqrt{2a r_1}$$

$$= \frac{\kappa + 1}{4\pi \mu} K_I \sqrt{2\pi r_1}$$

which is consistent with Eq. (3.45).

3.4.2 Mode II Crack

Consider a cracked plate of infinite extent that is subjected to uniform shear stress τ_0 at infinity as shown in Figure 3.5. This is a basic Mode II problem with the following skew-symetric boundary conditions:

$$\sigma_{xy} = 0, \, \sigma_{yy} = 0 \quad \text{at } |x| \le a \text{ and } y = 0 \quad \text{(crack surfaces)}$$
$$\sigma_{xx} = \sigma_{yy} = 0, \, \sigma_{xy} = \tau_0 \quad \text{at } x^2 + y^2 \to \infty \tag{3.48}$$

It can be shown that the following Westergaard function,

$$Z_{II}(z) = \frac{\tau_0 z}{\sqrt{z^2 - a^2}} \tag{3.49}$$

yields stresses that satisfy the boundary conditions Eq. (3.48) and, hence, yields the solution for the problem. The stress field can be computed according to Eqs. (3.34)

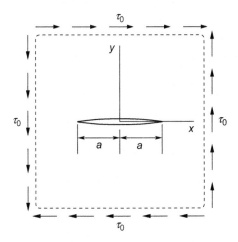

FIGURE 3.5

A crack in an infinite elastic plane subjected to pure in-plane shear.

and (3.49). Following the same procedure described for the Mode I crack and using the polar coordinates defined in Eq. (3.37) and Figure 3.4, we have the complete stress field,

$$
\sigma_{xx} = \frac{\tau_0 r}{\sqrt{r_1 r_2}} \left[2 \sin\left(\theta - \frac{1}{2}\theta_1 - \frac{1}{2}\theta_2 \right) - \frac{a^2}{r_1 r_2} \sin\theta \cos\frac{3}{2}(\theta_1 + \theta_2) \right]
$$

$$
\sigma_{yy} = \frac{\tau_0 a^2 r}{(r_1 r_2)^{3/2}} \sin\theta \cos\frac{3}{2}(\theta_1 + \theta_2) \tag{3.50}
$$

$$
\sigma_{xy} = \frac{\tau_0 r}{\sqrt{r_1 r_2}} \left[\cos\left(\theta - \frac{1}{2}\theta_1 - \frac{1}{2}\theta_2 \right) - \frac{a^2}{r_1 r_2} \sin\theta \sin\frac{3}{2}(\theta_1 + \theta_2) \right]
$$

Along the crack extended line ($\theta_1 = \theta_2 = \theta = 0$) and near the crack tip ($r_1/a \ll 1$), the stresses are

$$
\sigma_{yy} = 0
$$

$$
\sigma_{xy} = \frac{\tau_0 \sqrt{\pi a}}{\sqrt{2\pi r_1}}
$$

Comparing these stresses with Eq. (3.1) in Section 3.1, we have the Mode II stress intensity factor K_{II} for the crack problem as

$$
K_{II} = \tau_0 \sqrt{\pi a} \tag{3.51}
$$

and the Mode I stress intensity factor $K_I = 0$. In general, the stresses in the vicinity of the right crack tip are derived as

$$\sigma_{xx} = -\frac{K_{II}}{\sqrt{2\pi r}} \sin \frac{1}{2}\theta \left(2 + \cos \frac{\theta}{2} \cos \frac{3\theta}{2}\right)$$

$$\sigma_{yy} = \frac{K_{II}}{\sqrt{2\pi r}} \sin \frac{\theta}{2} \cos \frac{\theta}{2} \cos \frac{3}{2}\theta \qquad (3.52)$$

$$\sigma_{xy} = \frac{K_{II}}{\sqrt{2\pi r}} \cos \frac{1}{2}\theta \left(1 - \sin \frac{1}{2}\theta \sin \frac{3}{2}\theta\right)$$

and the near-tip displacements are

$$u_x = \frac{K_{II}}{8\mu\pi} \sqrt{2\pi r} \left[(2\kappa + 3)\sin \frac{\theta}{2} + \sin \frac{3\theta}{2}\right]$$

$$\qquad (3.53)$$

$$u_y = -\frac{K_{II}}{8\mu\pi} \sqrt{2\pi r} \left[(2\kappa - 3)\cos \frac{\theta}{2} + \cos \frac{3\theta}{2}\right]$$

where the origin of the (r,θ) system has been shifted to the right crack tip.

Equation (3.52) shows that stresses also have an inverse square root singularity at the crack tip and the intensity of the singular stress field is described by the Mode II stress intensity factor. It will be seen again in Section 3.7 that the near-tip stresses Eq. (3.52) and displacements Eq. (3.53) hold for any cracked body under Mode II deformation conditions with differences only in the value of K_{II}. Hence, once the stress σ_{xy} along the crack extended line is known, the Mode II stress intensity factor can be obtained from

$$K_{II} = \lim_{r \to 0} \sqrt{2\pi r}\,\sigma_{xy}(\theta = 0) \qquad (3.54)$$

The crack surface displacment may be obtained by setting $y = 0$ in Eq. (3.34) (B is ignored as it represents rigid displacements):

$$4\mu u_x = (\kappa + 1)\,\text{Im}\hat{Z}_{II}$$

$$4\mu u_y = (1 - \kappa)\,\text{Re}\hat{Z}_{II}$$

Since

$$\hat{Z}_{II} = \tau_0 \sqrt{z^2 - a^2}$$

we have

$$\hat{Z}_{II} = \tau_0 \sqrt{x^2 - a^2} \quad \text{at } y = 0$$

The crack surfaces lie in the region $|x| < a$. Thus,

$$\hat{Z}_{II} = i\tau_0 \sqrt{a^2 - x^2}$$

It is then obvious that the displacements of the upper crack surface are

$$u_y = 0$$

$$u_x = \frac{\kappa + 1}{4\mu} \tau_0 \sqrt{a^2 - x^2} \tag{3.55}$$

The preceding expressions show that the two crack surfaces slide with each other.

3.4.3 Mode III Crack

The Mode III fracture is associated with the anti-plane deformation for which the displacements are given by

$$u_x = 0, \quad u_y = 0, \quad u_z = w(x,y) \tag{3.56}$$

The nonvanishing strains are thus given by

$$e_{xz} = \frac{1}{2} \frac{\partial w}{\partial x}, \quad e_{yz} = \frac{1}{2} \frac{\partial w}{\partial y} \tag{3.57}$$

and the corresponding stresses follow Hooke's law:

$$\sigma_{xz} = 2\mu e_{xz}, \quad \sigma_{yz} = 2\mu e_{yz} \tag{3.58}$$

The equations of equilibrium reduce to

$$\frac{\partial \sigma_{xz}}{\partial x} + \frac{\partial \sigma_{yz}}{\partial y} = 0 \tag{3.59}$$

which can be written in terms of displacement w by using Eqs. (3.57) and (3.58) as follows:

$$\nabla^2 w = 0 \tag{3.60}$$

Thus, w must be a harmonic function. Let

$$w = \frac{1}{\mu} \text{Im}\{Z_{III}(z)\} \tag{3.61}$$

where $Z_{III}(z)$ is an analytic function. Substituting Eq. (3.61) in Eq. (3.57) and then Eq. (3.58), the stresses can be represented by $Z_{III}(z)$ as follows:

$$\sigma_{xz} - i\sigma_{yz} = -iZ'_{III}(z) \tag{3.62}$$

Now consider an infinite cracked body under anti-plane shear stress S shown in Figure 3.6. The boundary conditions of the crack problem are given by

$$\sigma_{yz} = 0 \quad \text{at } |x| \leq a \text{ and } y = 0$$
$$\sigma_{yz} = S \quad \text{at } |y| \to \infty \tag{3.63}$$

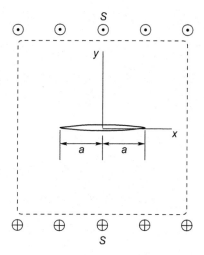

FIGURE 3.6

A crack in an infinite elastic body subjected to anti-plane shear.

Choose

$$Z_{III} = S\sqrt{z^2 - a^2} \tag{3.64}$$

We can easily show that the stresses calculated from this function satisfy the boundary conditions Eq. (3.63). Substituting Eq. (3.64) in Eq. (3.62) and using the polar coordinates defined in Eq. (3.37), the stress components can be obtained as

$$\sigma_{yz} = \text{Re}\{Z'_{III}\} = \frac{Sr}{\sqrt{r_1 r_2}} \cos\left(\theta - \frac{1}{2}\theta_1 - \frac{1}{2}\theta_2\right)$$

$$\sigma_{xz} = \text{Im}\{Z'_{III}\} = \frac{Sr}{\sqrt{r_1 r_2}} \sin\left(\theta - \frac{1}{2}\theta_1 - \frac{1}{2}\theta_2\right) \tag{3.65}$$

Again consider the stress along the crack extended line ($\theta_1 = \theta_2 = \theta = 0$) and near the crack tip ($r_1/a \ll 1$). It is obtained from Eq. (3.65) as

$$\sigma_{yz} = \frac{S\sqrt{\pi a}}{\sqrt{2\pi r_1}}$$

Comparing these stresses with Eq. (3.1) in Section 3.1, we have the Mode III stress intensity factor K_{III} for the crack problem as

$$K_{III} = S\sqrt{\pi a} \tag{3.66}$$

In general, the near-tip stresses are obtained as

$$\sigma_{yz} = \frac{K_{III}}{\sqrt{2\pi r_1}} \cos \frac{1}{2}\theta_1$$

$$\sigma_{xz} = -\frac{K_{III}}{\sqrt{2\pi r_1}} \sin \frac{1}{2}\theta_1 \tag{3.67}$$

and the anti-plane displacement is

$$w = \sqrt{\frac{2}{\pi}} \frac{K_{III}}{\mu} \sqrt{r_1} \sin \frac{1}{2}\theta_1 \tag{3.68}$$

The forms of Eqs. (3.67) and (3.68) hold for general Mode III cracks and the Mode III stress intensity factor is calculated from

$$K_{III} = \lim_{r_1 \to 0} \sqrt{2\pi r_1} \, \sigma_{yz}(\theta = 0) \tag{3.69}$$

The complete displacement field is obtained using Eqs. (3.64) and (3.61) as

$$w = u_z = \frac{S}{\mu} \sqrt{r_1 r_2} \sin \frac{1}{2}(\theta_1 + \theta_2)$$

On the upper crack surface ($y = 0^+$, $|x| < a$, or $\theta_1 = \pi$, $\theta_2 = 0$), the displacement is

$$u_z = \frac{S}{\mu} \sqrt{r_1 r_2} = \frac{S}{\mu} \sqrt{a^2 - x^2} \tag{3.70}$$

3.4.4 Complex Representation of Stress Intensity Factor

Stress intensity factor is a key concept in linear elastic fracture mechanics. It will be seen in Section 3.7 that the asymptotic stress and displacement fields near a crack tip have universal forms as described in Eqs. (3.44), (3.45), (3.52), and (3.53). Solving for the stresses and displacements around a crack tip thus reduces to finding the stress intensity factors, which may be directly calculated from the solutions of the complex potential functions. It follows from the near-tip stress field Eqs. (3.44) and (3.52) that (now use (r_1, θ_1) at the crack tip $z = a$)

$$\sigma_{xx} + \sigma_{yy} = \frac{2K_I}{\sqrt{2\pi r_1}} \cos \frac{1}{2}\theta_1 \qquad \text{for Mode I}$$

$$\sigma_{xx} + \sigma_{yy} = -\frac{2K_{II}}{\sqrt{2\pi r_1}} \sin \frac{1}{2}\theta_1 \qquad \text{for Mode II}$$

For combined loading, we have

$$\sigma_{xx} + \sigma_{yy} = \frac{2K_I}{\sqrt{2\pi r_1}} \cos \frac{1}{2}\theta_1 - \frac{2K_{II}}{\sqrt{2\pi r_1}} \sin \frac{1}{2}\theta_1$$

Defining the complex stress intensity factor K,

$$K = K_I - iK_{II} \tag{3.71}$$

and recalling the definition of polar coordinates (r_1, θ_1),

$$z - a = r_1 e^{i\theta_1}$$

one can show that

$$\sigma_{xx} + \sigma_{yy} = 2\,\mathrm{Re}\left\{ \frac{K}{\sqrt{2\pi(z-a)}} \right\} \quad z \to a$$

Since

$$\sigma_{xx} + \sigma_{yy} = 4\,\mathrm{Re}\{\psi'(z)\}$$

we obtain

$$\mathrm{Re}\left\{ \frac{K}{\sqrt{2\pi(z-a)}} \right\} = 2\,\mathrm{Re}\{\psi'(z)\} \quad z \to a$$

The complex stress intensity factor at the crack tip $z = a$ follows from this equation:

$$K = 2\sqrt{2\pi} \lim_{z \to a} \{ (\sqrt{z-a})\psi'(z) \} \tag{3.72}$$

In the last step of deriving the expression, we have used the relation

$$\mathrm{Re}\{f(z)\} = \mathrm{Re}\{g(z)\} \Rightarrow f(z) = g(z) + iC$$

where $f(z)$ and $g(z)$ are analytic functions and C is a real constant.

Proof

Let

$$h(z) = f(z) - g(z) = U_3 + iV_3$$

Thus, $h(z)$ is also analytic. Since $\mathrm{Re}\{h(z)\} = 0$, i.e.,

$$U_3 = 0$$

we have from the Cauchy-Riemann equation (3.11)

$$\frac{\partial V_3}{\partial x} = -\frac{\partial U_3}{\partial y} = 0$$

$$\frac{\partial V_3}{\partial y} = \frac{\partial U_3}{\partial x} = 0$$

Thus,

$$V_3 = \text{real constant} = C$$

i.e.,

$$f(z) = g(z) + iC$$

If the Westergaard functions are used, we can obtain the following expression for the complex stress intensity factor at the tip $z = a$:

$$K = \sqrt{2\pi} \lim_{z \to a} \left\{ \sqrt{z - a}(Z_I - iZ_{II}) \right\} \tag{3.73}$$

For Mode III cracks, it follows from Eqs. (3.62) and (3.67) that

$$K_{III} = \sqrt{2\pi} \lim_{z \to a} \left\{ \sqrt{z - a}Z'_{III}(z) \right\} \tag{3.74}$$

3.5 FUNDAMENTAL SOLUTIONS OF STRESS INTENSITY FACTOR

In the previous section, the stress intensity factors for a crack in an infinite plane were calculated under uniform loading conditions. In engineering applications, a cracked body is generally subjected to nonuniformly distributed loads. Stress intensity factors of a cracked body under arbitrary loading conditions may be obtained using the superposition method and the fundamental solutions, which are the stress intensity factors for a cracked body subjected to concentrated forces on the crack faces.

Consider a cracked body subjected to arbitrarily loads, as shown in Figure 3.7(a). The crack surfaces are assumed in a traction free state without loss of generality. The stresses and displacements in the cracked body can be obtained by superposing the corresponding solutions of the following two problems, as shown in Figures 3.7(b) and (c). The first problem is the same elastic body without the crack subjected to the same external loads. The solutions to this first problem can be obtained using conventional methods in the theory of elasticity because no cracks are present. The second problem consists of the same cracked body subjected to the crack face tractions, which are equal only in magnitude but opposite in sign to the tractions obtained at the same crack location in the first problem.

It is evident that the superposition of the solutions of the two problems satisfies all the boundary conditions and the traction free conditions at the crack surfaces. In linear elastic fracture mechanics, the solution to the first problem is assumed

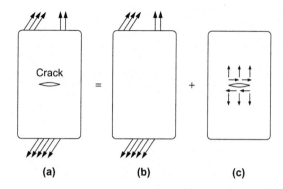

FIGURE 3.7

Superposition method of linear elastic crack problems (a) the original crack problem, (b) the first problem, (c) the second problem.

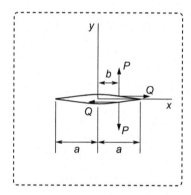

FIGURE 3.8

A crack in an infinite plate subjected to concentrated forces on the crack faces.

to be known and the stresses are finite at the location of the crack tip. The stress intensity factors therefore can be calculated from the solution to the second problem and may be obtained using the fundamental solutions. In the following sections, two fundamental solutions for a crack in an infinite plane are introduced.

3.5.1 A Finite Crack in an Infinite Plate

Consider an infinite plate containing a crack of length $2a$ subjected to a pair of compressive forces P per unit thickness at $x = b$, as shown in Figure 3.8 (the shear force Q is assumed to be absent for now). The boundary conditions of the crack problem are formulated as follows:

$$\sigma_{yy} = 0 \quad \text{at } |x| \le a, x \neq b \text{ and } y = 0$$

$$\int_{-a}^{a} \sigma_{yy} dx = -P \quad \text{at } y = 0^+ \text{ and } y = 0^-$$

(3.75)

$$\sigma_{xy} = 0 \quad \text{at } |x| \le a \text{ and } y = 0$$

$$\sigma_{xx}, \sigma_{yy}, \sigma_{xy} \to 0 \quad \text{at } x^2 + y^2 \to \infty$$

It can be shown that the Westergaard function

$$Z_I = \frac{P}{\pi(z-b)} \sqrt{\frac{a^2 - b^2}{z^2 - a^2}}$$

gives the stresses that satisfy the boundary conditions Eq. (3.75). Substituting the previous function into Eq. (3.73) yields the complex stress intensity factor at the right crack tip ($z = a$):

$$K_I - iK_{II} = \sqrt{2\pi} \lim_{z \to a} \left\{ \sqrt{z-a} \frac{P}{\pi(z-b)} \sqrt{\frac{a^2 - b^2}{z^2 - a^2}} \right\}$$

$$= \frac{P}{\sqrt{\pi a}} \sqrt{\frac{a+b}{a-b}}$$

Similarly, we can get the complex stress intensity factor at the left crack tip (note that $z - a$ in Eq. (3.73) should be replaced by $-(z+a)$ when evaluating the stress intensity factor at the tip $z = -a$). Finally, we have the stress intensity factors at the two crack tips:

$$K_I = \frac{P}{\sqrt{\pi a}} \sqrt{\frac{a+b}{a-b}}, \quad \text{at the right crack tip } (x = a)$$

$$= \frac{P}{\sqrt{\pi a}} \sqrt{\frac{a-b}{a+b}}, \quad \text{at the left crack tip } (x = -a)$$

(3.76)

When the crack faces are subjected to a pair of shearing forces Q per unit thickness at $x = b$, as shown in Figure 3.8 (now the compressive force P is assumed absent), the Westergaard function

$$Z_{II} = \frac{Q}{\pi(z-b)} \sqrt{\frac{a^2 - b^2}{z^2 - a^2}}$$

gives the stresses that satisfy the following boundary conditions of the Mode II crack problem, that is,

$$\sigma_{xy} = 0 \quad \text{at } |x| \le a, x \neq b \text{ and } y = 0$$

$$\int_{-a}^{a} \sigma_{xy} dx = -Q \quad \text{at } y = 0^+ \text{ and } y = 0^- \tag{3.77}$$

$$\sigma_{yy} = 0 \quad \text{at } |x| \le a \text{ and } y = 0$$

$$\sigma_{xx}, \sigma_{yy}, \sigma_{xy} \to 0 \quad \text{at } x^2 + y^2 \to \infty$$

Substituting the Westergaard function into Eq. (3.73), we have the Mode II stress intensity factors:

$$K_{II} = \frac{Q}{\sqrt{\pi a}} \sqrt{\frac{a+b}{a-b}}, \quad \text{at the right crack tip } (x = a)$$

$$= \frac{Q}{\sqrt{\pi a}} \sqrt{\frac{a-b}{a+b}}, \quad \text{at the left crack tip } (x = -a) \tag{3.78}$$

3.5.2 Stress Intensity Factors for a Crack Subjected to Arbitrary Crack Face Loads

In linear elastcity, it has been known that the stress (and displacement) in an elastic body subjected to a number of external loads is the sum of the stresses corresponding to each individual load. This is the so-called superposition principle of linear elastic systems. We now apply this principle to problems of cracks in linear elastic solids. The fundamental solutions Eqs. (3.76) and (3.78) can thus be used to obtain stress intensity factors under arbitrary crack face loads.

Consider a crack of length $2a$ in an infinite plate subjected to arbitrarily distributed pressure $p(x)$ on the crack faces. To obtain the stress intensity factor, we consider an infinitesimal length element $d\xi$ at $x = \xi$ on the crack face. The force exerted on this length element is $p(\xi)d\xi$, which induces the following stress intensity factors according to the fundamental solution Eq. (3.76):

$$dK_I = \frac{p(\xi)d\xi}{\sqrt{\pi a}} \sqrt{\frac{a+\xi}{a-\xi}}, \quad \text{at the right crack tip } (x = a)$$

$$= \frac{p(\xi)d\xi}{\sqrt{\pi a}} \sqrt{\frac{a-\xi}{a+\xi}}, \quad \text{at the left crack tip } (x = -a)$$

The total stress intensity factors due to the distributed pressure can be obtained by integrating the preceding expression along the crack face (from $\xi = -a$ to a).

We have

$$K_I = \frac{1}{\sqrt{\pi a}} \int\limits_{-a}^{a} p(\xi) \sqrt{\frac{a+\xi}{a-\xi}} \, d\xi, \quad \text{at the right crack tip } (x = a)$$

$$= \frac{1}{\sqrt{\pi a}} \int\limits_{-a}^{a} p(\xi) \sqrt{\frac{a-\xi}{a+\xi}} \, d\xi, \quad \text{at the left crack tip } (x = -a) \qquad (3.79)$$

Similarly, the Mode II stress intensity factors corresponding to the distributed shearing traction $q(x)$ on the crack surfaces can be obtained by using the fundamental solution Eq. (3.78) as follows:

$$K_{II} = \frac{1}{\sqrt{\pi a}} \int\limits_{-a}^{a} q(\xi) \sqrt{\frac{a+\xi}{a-\xi}} \, d\xi, \quad \text{at the right crack tip } (x = a)$$

$$= \frac{1}{\sqrt{\pi a}} \int\limits_{-a}^{a} q(\xi) \sqrt{\frac{a-\xi}{a+\xi}} \, d\xi, \quad \text{at the left crack tip } (x = -a) \qquad (3.80)$$

3.5.3 A Semi-infinite Crack in an Infinite Medium

Consider an infinite medium containing a semi-infinite crack subjected to a pair of compressive forces P per unit thickness, as shown in Figure 3.9. The boundary

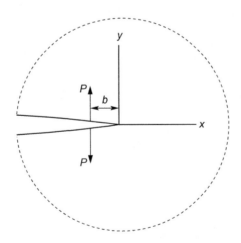

FIGURE 3.9

A semi-infinite crack in an infinite plate subjected to concentrated forces on the crack faces.

conditions of the crack problem are

$$\sigma_{yy} = 0 \quad \text{at} \ -\infty < x < 0, \ x \neq -b \ \text{and} \ y = 0$$

$$\int_{-\infty}^{0} \sigma_{yy} dx = -P \quad \text{at} \ y = 0^+ \ \text{and} \ y = 0^-$$

$$\sigma_{xy} = 0 \quad \text{at} \ -\infty < x < 0 \ \text{and} \ y = 0$$

$$\sigma_{xx}, \sigma_{yy}, \sigma_{xy} \to 0 \quad \text{at} \ x^2 + y^2 \to \infty$$

The Westergaard function for the semi-infinite crack problem is given by

$$Z_I = \frac{P}{\pi(z+b)} \sqrt{\frac{b}{z}}$$

Substituting the function here into Eq. (3.73), we have the Mode I stress intensity factor at the crack tip ($z = 0$):

$$K_I = \sqrt{2\pi} \lim_{z \to 0} \left\{ \sqrt{z} \frac{P}{\pi(z+b)} \sqrt{\frac{b}{z}} \right\}$$

$$= P \sqrt{\frac{2}{\pi b}} \tag{3.81}$$

3.6 FINITE SPECIMEN SIZE EFFECTS

In Sections 3.4 and 3.5, we have introduced stress intensity factor solutions for some basic crack problems of infinite media. These solutions may be used when the crack length is much smaller than the in-plane size of the cracked body. When the crack length is of the order of the size of the cracked body, the stress intensity factor will be influenced by the size of the cracked body. In general, numerical methods (for example, the finite element method) are used to calculate stress intensity factors for cracks in finite size media.

Now consider a center-cracked plate subjected to tension as shown in Figure 3.10. The height and width of the plate are $2H$ and $2b$, respectively. The crack length is $2a$. The general form of the stress intensity factor may be expressed as

$$K_I = \sigma \sqrt{\pi a} F(a/b, a/H) \tag{3.82}$$

where $F(a/b, a/H)$ is a dimensionless function of a/b and a/H. When $H \gg b$, the plate may be regarded as an infinite strip. In this case, the parameter a/H may be dropped in $F(a/b, a/H)$ in Eq. (3.82) and some empirical expressions for $F(a/b)$

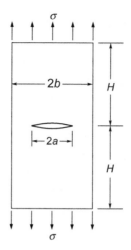

FIGURE 3.10

A finite plate with a center crack subjected to tension.

were proposed by curve-fitting the numerical result of the stress intensity factor. For example, Irwin [3-1] proposed the following approximate formula:

$$F(a/b) = 1 + 0.128 \left(\frac{a}{b}\right) - 0.288 \left(\frac{a}{b}\right)^2 + 1.525 \left(\frac{a}{b}\right)^3 \tag{3.83}$$

The relative error between this formula here and that of the numerical solution is less that 0.5% when $a/b \leq 0.7$. It is noted that the preceding stress intensity factor for the center-cracked plate reduces to that for an infinite plate ($\sigma \sqrt{\pi a}$) when $a/b \rightarrow 0$.

The Appendix lists stress intensity factor solutions for some typical cracked specimens.

3.7 WILLIAMS' CRACK TIP FIELDS

In the previous sections, explicit stress and displacement fields near the crack tip were obtained using the Westergaard function method. One common characteristic feature of these solutions is that the stresses have an inverse square root singularity at the crack tip and the unambiguous functional forms of these near-tip stress and displacement fields do not depend on the applied load and the geometry of the cracked body. For instance, the near-tip stress and displacement fields for a crack in an infinite plate subjected to remote loading are exactly the same as those for the cracked plate subjected to crack face concentrated loading. This observation regarding the unique functional form for the near-tip fields can be verified using the eigenfunction expansion method for any crack geometries and loading conditions presented by Williams [3-5, 3-6].

3.7.1 Williams' Crack Tip Stress and Displacement Fields: Mode I and II

It has been shown in Section 3.2 that stresses and displacements can be repesented by the Airy stress function ϕ, which satisfies the following biharmonic equation:

$$\nabla^2 \nabla^2 \phi = 0 \tag{3.84}$$

where the Laplacian operator ∇^2 in the polar coordinate system (r,θ) centered at the crack tip (see Figure 3.11) is given by

$$\nabla^2 = \frac{\partial^2}{\partial r^2} + \frac{1}{r}\frac{\partial}{\partial r} + \frac{1}{r^2}\frac{\partial^2}{\partial \theta^2}$$

The Airy stress function near the crack tip may be expanded into the following series:

$$\phi = \sum_{n=0}^{\infty} r^{\lambda_n+1} F_n(\theta) \tag{3.85}$$

where λ_n are eigenvalues to be determined and $F_n(\theta)$ are the corresponding eigenfunctions. Substitution of Eq. (3.85) in Eq. (3.84) yields

$$\frac{d^4 F_n(\theta)}{d\theta^4} + 2(\lambda_n^2 + 1)\frac{d^2 F_n(\theta)}{d\theta^2} + (\lambda_n^2 - 1)^2 F_n(\theta) = 0$$

The solution of this equation for $F_n(\theta)$ has the following form:

$$\begin{aligned} F_n(\theta) = A_n \sin(\lambda_n + 1)\theta + B_n \cos(\lambda_n + 1)\theta \\ + C_n \sin(\lambda_n - 1)\theta + D_n \cos(\lambda_n - 1)\theta \end{aligned} \tag{3.86}$$

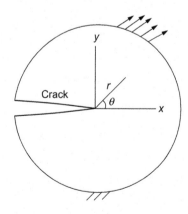

FIGURE 3.11

A cracked plate and the coordinate systems.

where A_n, B_n, C_n, and D_n are unknown constants. The stresses need to satisfy the following traction free boundary conditions along the crack surfaces:

$$\sigma_{\theta\theta} = \sigma_{r\theta} = 0, \quad \theta = \pm\pi \tag{3.87}$$

Mode I Case

For Mode I crack problems, the Airy stress function is an even function of θ. The constants A_n and C_n in Eq. (3.86) become zero and the Airy function can be written as

$$\phi = \sum_{n=0}^{\infty} r^{\lambda_n+1} [B_n \cos(\lambda_n + 1)\theta + D_n \cos(\lambda_n - 1)\theta] \tag{3.88}$$

The stresses in the polar coordinate system can be obtained from the Airy function as follows:

$$\sigma_{rr} = \frac{1}{r^2}\frac{\partial^2\phi}{\partial\theta^2} + \frac{1}{r}\frac{\partial\phi}{\partial r} = \sum_{n=0}^{\infty} r^{\lambda_n-1} \left[F_n''(\theta) + (\lambda_n + 1) F_n(\theta) \right]$$

$$= -\sum_{n=0}^{\infty} \lambda_n r^{\lambda_n-1} [B_n(\lambda_n + 1)\cos(\lambda_n + 1)\theta + D_n(\lambda_n - 3)\cos(\lambda_n - 1)\theta]$$

$$\sigma_{\theta\theta} = \frac{\partial^2\phi}{\partial r^2} = \sum_{n=0}^{\infty} r^{\lambda_n-1} \lambda_n (\lambda_n + 1) F_n(\theta)$$

$$= \sum_{n=0}^{\infty} r^{\lambda_n-1} \lambda_n (\lambda_n + 1) [B_n \cos(\lambda_n + 1)\theta + D_n \cos(\lambda_n - 1)\theta]$$

$$\sigma_{r\theta} = -\frac{\partial}{\partial r}\left(\frac{1}{r}\frac{\partial\phi}{\partial\theta}\right) = -\sum_{n=0}^{\infty} \lambda_n r^{\lambda_n-1} F_n'(\theta)$$

$$= \sum_{n=0}^{\infty} \lambda_n r^{\lambda_n-1} [B_n(\lambda_n + 1)\sin(\lambda_n + 1)\theta + D_n(\lambda_n - 1)\sin(\lambda_n - 1)\theta] \tag{3.89}$$

The displacements in the polar coordinate system may be obtained using Hooke's law and the strain-displacement relations given as:

$$\frac{\partial u_r}{\partial r} = e_{rr} = \frac{1}{2\mu}\left[\sigma_{rr} - \frac{3-\kappa}{4}(\sigma_{rr} + \sigma_{\theta\theta})\right]$$

$$\frac{u_r}{r} + \frac{1}{r}\frac{\partial u_\theta}{\partial\theta} = e_{\theta\theta} = \frac{1}{2\mu}\left[\sigma_{\theta\theta} - \frac{3-\kappa}{4}(\sigma_{rr} + \sigma_{\theta\theta})\right] \tag{3.90}$$

$$\frac{1}{r}\frac{\partial u_r}{\partial\theta} + r\frac{\partial}{\partial r}\left(\frac{u_\theta}{r}\right) = e_{r\theta} = \frac{1}{2\mu}\sigma_{r\theta}$$

where

$$\kappa = \begin{cases} 3 - 4v & \text{for plane strain} \\ \dfrac{3 - v}{1 + v} & \text{for plane stress} \end{cases}$$

Substituting the stresses Eq. (3.89) in Eq. (3.90) and performing appropriate integration gives the following series form solutions for the displacements:

$$u_r = \frac{1}{2\mu} \sum_{n=0}^{\infty} r^{\lambda_n} \{ -(\lambda_n + 1)F_n(\theta) + (1 + \kappa)D_n \cos(\lambda_n - 1)\theta \}$$

$$u_\theta = \frac{1}{2\mu} \sum_{n=0}^{\infty} r^{\lambda_n} \{ -F_n'(\theta) + (1 + \kappa)D_n \sin(\lambda_n - 1)\theta \}$$

(3.91)

Substituting the stresses Eq. (3.89) in the boundary conditions Eq. (3.87) leads to the following two simultaneous equations for the constants B_n and D_n:

$$B_n \cos(\lambda_n + 1)\pi + D_n \cos(\lambda_n - 1)\pi = 0$$

$$B_n(\lambda_n + 1)\sin(\lambda_n + 1)\pi + D_n(\lambda_n - 1)\sin(\lambda_n - 1)\pi = 0$$

(3.92)

The existence of nontrivial solutions for B_n and D_n for the system of homogeneous equations leads to the following characteristic equation of the eigenvalue λ_n:

$$\sin(2\lambda_n \pi) = 0 \tag{3.93}$$

The roots of this equation are

$$\lambda_n = \frac{n}{2}, \quad n = 0, \pm 1, \pm 2, \dots$$

It follows from the displacement expression Eq. (3.91) that the negative values of λ_n would give rise to infinite displacements at the crack tip $r = 0$, which is not physically permissible, and hence should be excluded. A zero λ_n ($n = 0$) leads to physically impermissible unbounded strain energy in a small disc area around the crack tip and should also be excluded. Hence, only positive eigenvalues are kept. Thus,

$$\lambda_n = \frac{n}{2}, \quad n = 1, 2, \dots \tag{3.94}$$

Substituting the eigenvalues above in Eq. (3.92), we obtain the following relation between B_n and D_n:

$$B_n = -\frac{(\frac{n}{2} - 1)\sin(\frac{n}{2} - 1)\pi}{(\frac{n}{2} + 1)\sin(\frac{n}{2} + 1)\pi} D_n = -\frac{n - 2}{n + 2} D_n, \quad n = 1, 3, 5, \dots$$

$$B_n = -\frac{\cos\left(\frac{n}{2} - 1\right)\pi}{\cos\left(\frac{n}{2} + 1\right)\pi} D_n = -D_n, \quad n = 2, 4, 6, \dots$$

(3.95)

The constants D_n depend on the loading and boundary conditions of the specific crack problem as only the traction-free conditions on the crack faces are used in deriving the crack tip asymptotic solutions. Substituting the eigenvalues Eq. (3.94) in Eq. (3.89), we obtain the following series form solution for Mode I stresses near the crack tip:

$$\sigma_{rr} = -\sum_{n=1}^{\infty} r^{\frac{n}{2}-1} \left(\frac{n}{2}\right) \left[B_n \left(\frac{n}{2}+1\right) \cos\left(\frac{n}{2}+1\right)\theta + D_n \left(\frac{n}{2}-3\right) \cos\left(\frac{n}{2}-1\right)\theta \right]$$

$$\sigma_{\theta\theta} = \sum_{n=1}^{\infty} r^{\frac{n}{2}-1} \left(\frac{n}{2}\right) \left(\frac{n}{2}+1\right) \left[B_n \cos\left(\frac{n}{2}+1\right)\theta + D_n \cos\left(\frac{n}{2}-1\right)\theta \right] \qquad (3.96)$$

$$\sigma_{r\theta} = \sum_{n=1}^{\infty} r^{\frac{n}{2}-1} \left(\frac{n}{2}\right) \left[B_n \left(\frac{n}{2}+1\right) \sin\left(\frac{n}{2}+1\right)\theta + D_n \left(\frac{n}{2}-1\right) \sin\left(\frac{n}{2}-1\right)\theta \right]$$

where B_n and D_n satisfy relation Eq. (3.95).

The first two terms ($n = 1, 2$) in the series solution just shown are given by

$$\sigma_{rr} = D_1 r^{-1/2} \left[-\frac{1}{4} \cos\frac{3\theta}{2} + \frac{5}{4} \cos\frac{\theta}{2} \right] + 2D_2 \cos 2\theta + 2D_2 + O\left(r^{1/2}\right)$$

$$\sigma_{\theta\theta} = D_1 r^{-1/2} \left[\frac{1}{4} \cos\frac{3\theta}{2} + \frac{3}{4} \cos\frac{\theta}{2} \right] - 2D_2 \cos 2\theta + 2D_2 + O\left(r^{1/2}\right) \qquad (3.97)$$

$$\sigma_{r\theta} = D_1 r^{-1/2} \left[\frac{1}{4} \sin\frac{3\theta}{2} + \frac{1}{4} \sin\frac{\theta}{2} \right] - 2D_2 \sin 2\theta + O\left(r^{1/2}\right)$$

Clearly only the first terms ($n = 1$) are singular at the crack tip. With the definition of Mode I stress intensity factor,

$$K_I = \lim_{r \to 0} \sqrt{2\pi r} \sigma_{\theta\theta}(r, 0)$$

the constant D_1 in the first terms is related to K_I as follows:

$$D_1 = \frac{K_I}{\sqrt{2\pi}} \qquad (3.98)$$

The constant D_2 in the second terms is related to the so-called T-stress. The rectangular stress components corresponding to the stresses in Eq. (3.97) are

$$\sigma_{xx} = D_1 r^{-1/2} \cos\frac{1}{2}\theta \left(1 - \sin\frac{1}{2}\theta \sin\frac{3}{2}\theta\right) + T + O\left(r^{1/2}\right)$$

$$\sigma_{yy} = D_1 r^{-1/2} \cos\frac{1}{2}\theta \left(1 + \sin\frac{1}{2}\theta \sin\frac{3}{2}\theta\right) + O\left(r^{1/2}\right) \qquad (3.99)$$

$$\sigma_{xy} = D_1 r^{-1/2} \sin\frac{1}{2}\theta \cos\frac{1}{2}\theta \cos\frac{3}{2}\theta + O\left(r^{1/2}\right)$$

where $T = 4D_2$ is the T-stress, the constant term in the σ_{xx} expansion. It has been shown that the T-stress significantly influences the shape and size of the plastic zone around the crack tip [3-7].

It is noted that the first three terms in the expansion of the opening stress ahead of the crack tip $(x > 0, y = 0)$ are

$$\sigma_{yy} = \frac{K_I}{\sqrt{2\pi x}} + b_0 x^{1/2} + b_1 x^{3/2}$$

It is clear that there is no equivalent T-stress (constant stress term) in the crack opening stress expansion.

Mode II Case

For Mode II crack problems, the Airy stress function is an odd function of θ. The constants B_n and D_n in Eq. (3.86) become zero and the Airy function can be written as

$$\phi = \sum_{n=0}^{\infty} r^{\lambda_n+1} [A_n \sin(\lambda_n + 1)\theta + C_n \sin(\lambda_n - 1)\theta] \qquad (3.100)$$

The stresses in the polar coordinate system can be obtained from the Airy function as follows:

$$\sigma_{rr} = -\sum_{n=0}^{\infty} \lambda_n r^{\lambda_n-1} [A_n(\lambda_n + 1)\sin(\lambda_n + 1)\theta + C_n(\lambda_n - 3)\sin(\lambda_n - 1)\theta]$$

$$\sigma_{\theta\theta} = \sum_{n=0}^{\infty} r^{\lambda_n-1} \lambda_n(\lambda_n + 1)[A_n \sin(\lambda_n + 1)\theta + C_n \sin(\lambda_n - 1)\theta] \qquad (3.101)$$

$$\sigma_{r\theta} = -\sum_{n=0}^{\infty} \lambda_n r^{\lambda_n-1} [A_n(\lambda_n + 1)\cos(\lambda_n + 1)\theta + C_n(\lambda_n - 1)\cos(\lambda_n - 1)\theta]$$

The displacements in the polar coordinate system are given by

$$u_r = \frac{1}{2\mu} \sum_{n=0}^{\infty} r^{\lambda_n} \{-(\lambda_n + 1)F_n(\theta) + (1 + \kappa)C_n \sin(\lambda_n - 1)\theta\}$$

$$\qquad (3.102)$$

$$u_\theta = \frac{1}{2\mu} \sum_{n=0}^{\infty} r^{\lambda_n} \{-F'_n(\theta) - (1 + \kappa)C_n \cos(\lambda_n - 1)\theta\}$$

Substiting the stresses Eq. (3.101) into the boundary conditions Eq. (3.87) leads to the following two simultaneous equations for the constants A_n and C_n:

$$A_n \sin(\lambda_n + 1)\pi + C_n \sin(\lambda_n - 1)\pi = 0$$

$$\qquad (3.103)$$

$$A_n(\lambda_n + 1)\cos(\lambda_n + 1)\pi + C_n(\lambda_n - 1)\cos(\lambda_n - 1)\pi = 0$$

The existence of nontrivial solutions for A_n and C_n for the system of homogeneous equations here leads to the same characteristic equation (3.93). Hence, the eigenvalues λ_n are the same as those for Mode I in Eq. (3.94). Substituting the eigenvalues in Eq. (3.103) yields the following relation between A_n and C_n:

$$A_n = -\frac{\sin\left(\frac{n}{2}-1\right)\pi}{\sin\left(\frac{n}{2}+1\right)\pi}C_n = -C_n, \quad n = 1,3,5,\ldots$$

$$A_n = -\frac{\left(\frac{n}{2}-1\right)\cos\left(\frac{n}{2}-1\right)\pi}{\left(\frac{n}{2}+1\right)\cos\left(\frac{n}{2}+1\right)\pi}C_n = -\frac{n-2}{n+2}C_n, \quad n = 2,4,6,\ldots$$

(3.104)

Again, the constants C_n need to be determined using the loading and boundary conditions of the specific crack problem. Substitution of the eigenvalues Eq. (3.94) in Eq. (3.101) yields the following series form solution for Mode II stresses near the crack tip:

$$\sigma_{rr} = -\sum_{n=1}^{\infty} r^{\frac{n}{2}-1}\left(\frac{n}{2}\right)\left[A_n\left(\frac{n}{2}+1\right)\sin\left(\frac{n}{2}+1\right)\theta + C_n\left(\frac{n}{2}-3\right)\sin\left(\frac{n}{2}-1\right)\theta\right]$$

$$\sigma_{\theta\theta} = \sum_{n=1}^{\infty} r^{\frac{n}{2}-1}\left(\frac{n}{2}\right)\left(\frac{n}{2}+1\right)\left[A_n\sin\left(\frac{n}{2}+1\right)\theta + C_n\sin\left(\frac{n}{2}-1\right)\theta\right] \quad (3.105)$$

$$\sigma_{r\theta} = -\sum_{n=1}^{\infty} r^{\frac{n}{2}-1}\left(\frac{n}{2}\right)\left[A_n\left(\frac{n}{2}+1\right)\cos\left(\frac{n}{2}+1\right)\theta + C_n\left(\frac{n}{2}-1\right)\cos\left(\frac{n}{2}-1\right)\theta\right]$$

where A_n and C_n satisfy the relation Eq. (3.104). Again, only the first terms ($n = 1$) in this series solution are singular at the crack tip and they are given by

$$\sigma_{rr} = C_1 r^{-1/2}\left[\frac{3}{4}\sin\frac{3\theta}{2} - \frac{5}{4}\sin\frac{\theta}{2}\right]$$

$$\sigma_{\theta\theta} = C_1 r^{-1/2}\left[-\frac{3}{4}\sin\frac{3\theta}{2} - \frac{3}{4}\sin\frac{\theta}{2}\right] \quad (3.106)$$

$$\sigma_{r\theta} = C_1 r^{-1/2}\left[\frac{3}{4}\cos\frac{3\theta}{2} + \frac{1}{4}\cos\frac{\theta}{2}\right]$$

Using the definition of Mode II stress intensity factor,

$$K_{II} = \lim_{r\to 0} \sqrt{2\pi r}\,\sigma_{r\theta}(r,0)$$

the constant C_1 in the preceding equations can be related to K_{II} as follows:

$$C_1 = \frac{K_{II}}{\sqrt{2\pi}} \quad (3.107)$$

For combined Mode I and II crack problems, the stresses are the superposition of Eq. (3.96) for Mode I and Eq. (3.105) for Mode II. The first terms in the stress solutions, when transformed to the Cartesian components, have the same form as those in Eqs. (3.44) and (3.52) in Section 3.4 for the specific crack problems considered.

3.7.2 Williams' Crack Tip Stress and Displacement Fields: Mode III

The asymptotic series expansion method introduced for Mode I and Mode II cracks may also be used to obtain the crack tip fields for Mode III cracks. Corresponding to the basic equations (3.57), (3.58), and (3.59) in the Cartesian coordinates, the basic equations in the polar coordinates are given as follows:

$$e_{rz} = \frac{1}{2}\frac{\partial w}{\partial r}, \quad e_{\theta z} = \frac{1}{2r}\frac{\partial w}{\partial \theta} \tag{3.108}$$

$$\sigma_{rz} = 2\mu e_{rz}, \quad \sigma_{\theta z} = 2\mu e_{\theta z} \tag{3.109}$$

$$\frac{\partial \sigma_{rz}}{\partial r} + \frac{1}{r}\frac{\partial \sigma_{\theta z}}{\partial \theta} + \frac{\sigma_{rz}}{r} = 0 \tag{3.110}$$

The anti-plane displacement $w = w(r,\theta)$ satisfies the harmonic equation

$$\nabla^2 w = 0 \tag{3.111}$$

where ∇^2 in polar coordinates is given by

$$\nabla^2 = \frac{1}{r}\frac{\partial}{\partial r}\left(r\frac{\partial}{\partial r}\right) + \frac{1}{r^2}\frac{\partial^2}{\partial \theta^2}$$

The traction-free condition on the crack surfaces and the antisymmetry condition along the crack extended line lead to the boundary conditions:

$$\sigma_{\theta z} = 0 \quad \text{at } \theta = \pm\pi,\ r > 0$$
$$w = 0 \quad \text{at } \theta = 0,\ r > 0 \tag{3.112}$$

By the separation of variables approach, we assume the following expansion of displacement w near the crack tip ($r = 0$):

$$w(r,\theta) = \sum_{n=0}^{\infty} r^{\lambda_n} f_n(\theta) \tag{3.113}$$

where $f_n(\theta)$ are the eigenfunctions and λ_n are the eigenvalues to be determined. Substitution of Eq. (3.113) in Eq. (3.111) yields

$$\sum_{n=0}^{\infty} r^{\lambda_n - 2}\left(\lambda_n^2 f_n + f_n''\right) = 0$$

which leads to

$$f_n'' + \lambda_n^2 f_n = 0$$

The general solution for the preceding ordinary differential equation is readily obtained as

$$f_n(\theta) = A_n \cos \lambda_n \theta + B_n \sin \lambda_n \theta$$

Thus,

$$w(r,\theta) = \sum_{n=0}^{\infty} r^{\lambda_n} (A_n \cos \lambda_n \theta + B_n \sin \lambda_n \theta)$$

$$\sigma_{rz} = \mu \sum_{n=0}^{\infty} \lambda_n r^{\lambda_n-1} (A_n \cos \lambda_n \theta + B_n \sin \lambda_n \theta) \qquad (3.114)$$

$$\sigma_{\theta z} = \mu \sum_{n=0}^{\infty} \lambda_n r^{\lambda_n-1} (-A_n \sin \lambda_n \theta + B_n \cos \lambda_n \theta)$$

Substitition of Eq. (3.114) in the boundary conditions Eq. (3.112) gives

$$-A_n \sin \lambda_n \pi + B_n \cos \lambda_n \pi = 0$$

$$A_n = 0$$

For a nontrivial solution (B_n cannot be zero) for the previous homogeneous equations, we require

$$\cos \lambda_n \pi = 0$$

which yields

$$\lambda_n = n - \frac{1}{2}, \quad n = 1, 2, 3, ...$$

The negative eigenvalues have been excluded to avoid unbounded displacement at $r = 0$. Substituting this eigenvalues into Eq. (3.114) yields

$$w(r,\theta) = \sum_{n=1}^{\infty} B_n r^{n-\frac{1}{2}} \sin\left(n - \frac{1}{2}\right)\theta$$

$$\sigma_{rz} = \mu \sum_{n=1}^{\infty} \left(n - \frac{1}{2}\right) B_n r^{n-\frac{3}{2}} \sin\left(n - \frac{1}{2}\right)\theta \qquad (3.115)$$

$$\sigma_{\theta z} = \mu \sum_{n=1}^{\infty} \left(n - \frac{1}{2}\right) B_n r^{n-\frac{3}{2}} \cos\left(n - \frac{1}{2}\right)\theta$$

It is obvious from these expansions that only the first term ($n = 1$) produces singular stresses in the form

$$\sigma_{rz} = B_1 \frac{\mu}{2} r^{-\frac{1}{2}} \sin \frac{\theta}{2}$$

$$\sigma_{\theta z} = B_1 \frac{\mu}{2} r^{-\frac{1}{2}} \cos \frac{\theta}{2}$$

and the corresponding displacement is

$$w = B_1 r^{\frac{1}{2}} \sin \frac{\theta}{2}$$

At $\theta = 0$,

$$\sigma_{yz} = \sigma_{\theta z} = \frac{B_1 \mu}{2\sqrt{r}}$$

Thus, B_1 is related to the Mode III stress intensity factor K_{III} by

$$K_{III} = \sqrt{\frac{\pi}{2}} \, \mu B_1$$

and the near-tip stress and displacement field can be expressed in the form

$$\sigma_{rz} = \frac{K_{III}}{\sqrt{2\pi r}} \sin \frac{\theta}{2}$$

$$\sigma_{\theta z} = \frac{K_{III}}{\sqrt{2\pi r}} \cos \frac{\theta}{2}$$

$$w = \frac{K_{III}}{\pi \mu} \sqrt{2\pi r} \sin \frac{\theta}{2}$$

The stress components in the Cartesian coordinate system are

$$\sigma_{xz} = -\frac{K_{III}}{\sqrt{2\pi r}} \sin \frac{\theta}{2}$$

$$\sigma_{yz} = \frac{K_{III}}{\sqrt{2\pi r}} \cos \frac{\theta}{2}$$

$$w = \frac{K_{III}}{\pi \mu} \sqrt{2\pi r} \sin \frac{\theta}{2}$$

which are identical to Eqs. (3.67) and (3.68).

3.8 *K*-DOMINANCE

It is now clear that the stress field near a crack tip consists of two parts, one is singular at the crack tip and the other is nonsigular. The singular part becomes dominant if the location of interest approaches sufficiently close to the crack tip. In that case, the near-tip stress field can be effectively described by the singular stress field or, equivalently, by the stress intensity factor. This feature is of fundamental importance and is the foundation of the stress intensity factor-based fracture criterion of Irwin [3-1]. The successful use of the stress intensity factor to predict crack growth (fracture) depends on whether the singular stress term alone can represent the state of stress in the "fracture process zone," a region ahead of the crack tip in which the material failure process initiates. The knowledge of the *K*-dominance zone size is thus pertinent to the proper understanding and applicability of the fracture criterion.

To study the *K*-dominance zone size, the nonsingular part of the stress field is needed. For Mode I cracks, the normal stress σ_{yy} along the crack line ($x > 0$, $y = 0$) plays an essential role in crack extension. We define the degree of *K*-dominance at a point along the *x*-axis as

$$\Lambda = \frac{K_I/\sqrt{2\pi x}}{K_I/\sqrt{2\pi x} + \left|\text{nonsingular part of } \sigma_{yy}\right|} \tag{3.116}$$

where σ_{yy} is the total stress. Note that the absolute value of the nonsingular part of the normal stress is taken to ensure that the value of Λ does not exceed unity. A value of Λ close to unity means that the singular stress is dominant.

First consider a crack of length $2a$ in an infinite plate subjected to biaxial tension σ_0 as shown in Figure 3.3. The complete solution is available for this problem. From Eq. (3.41), the opening stress σ_{yy} along the extended crack line can be written as (now $x = 0$ is at the right crack tip)

$$\sigma_{yy} = \frac{(x+a)\sigma_0}{\sqrt{x(x+2a)}}, \quad x > 0$$

Since σ_{yy} is positive, the degree of *K*-dominance according to Eq. (3.116) is

$$\Lambda = \frac{K_I/\sqrt{2\pi x}}{\sigma_{yy}} = \frac{\sqrt{2+x/a}}{\sqrt{2}(1+x/a)}$$

The *K*-dominance zone size x_K for $\Lambda = 0.95$ can be determined from this equation with the result

$$x_K \approx 0.07a$$

The previous result indicates that, for a central crack in an infinite plate subjected to uniform biaxial tension, the singular stress consititutes at least 95% of the opening stress within the range $x \leq x_K$.

Next, consider the example of a finite rectangular plate containing a center crack subjected to a uniform tensile stress as shown in Figure 3.12. The normal stress σ_{yy} along the crack line ($x > 0$, $y = 0$) is calculated using the finite element method with $W/a = 2$. Figure 3.13 shows the normalized normal stress distributions $\sigma_{yy}\sqrt{2\pi x}/K_0$ for three plates of different heights, where $K_0 = \sigma_0\sqrt{\pi a}$. It is noted that, if the singular stress is dominant near the crack tip, the plot should be nearly a straight line in the

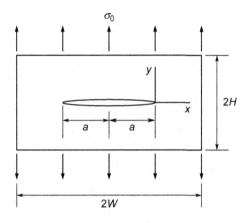

FIGURE 3.12

A cracked rectangular plate subjected to uniform tension.

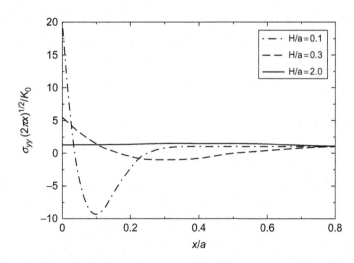

FIGURE 3.13

Normalized normal stress, $\sigma_{yy}\sqrt{2\pi x}/K_0$, along the crack line.

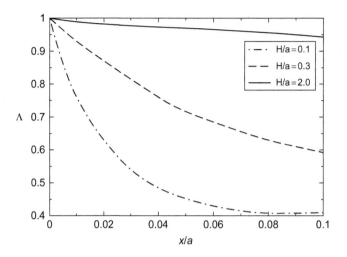

FIGURE 3.14

The degree of K-dominance, Λ, along the crack line.

crack tip region as in the case of $H/a = 2$. Figure 3.14 shows the plots of the degree of K-dominance, Λ, along the x-axis for the three cases.

The results in Figures 3.13 and 3.14 clearly show that, as the height of the plate decreases, the size of the K-dominance region also decreases. For $H/a = 0.1$ and $\Lambda = 0.95$, the size of the K-dominance zone can be estimated from Figure 3.14 to be $x_K \approx 0.002a$. In other words, the region in which the singular stress is dominant is very small and, thus, K may not be able to account for the entire fracture driving force. This is one of the reasons why the fracture toughness of a material measured in stress intensity factor may be influenced by the geometry of the test specimen and the loading condition [3-8].

3.9 IRWIN'S *K*-BASED FRACTURE CRITERION

From the continuum mechanics point of view, fracture is governed by the local stress and deformation conditions around the crack tip. It follows from the discussion in the last section that the near-tip stress field in the K-dominance zone can be represented by the inverse square-root singular stresses, which have universal functional forms in a local coordinate system centered at the crack tip. The geometries of cracked bodies and the loading conditions influence the crack tip singular field only through K ($K_I, K_{II},$ or K_{III} for different fracture modes), the stress intensity factor. Based on this fact, Irwin [3-1] proposed a fracture criterion, which states that crack growth occurs when the stress intensity factor reaches a critical value. Under Mode I conditions,

Irwin's criterion can be expressed as

$$K_I = K_c \tag{3.117}$$

where K_I is the applied stress intensity factor, which depends on the load level and the geometry of the cracked body, and K_c is the critical value of K_I at fracture, or fracture toughness.

Irwin's *K*-based fracture criterion differs from Griffith's free surface energy criterion in that the former focuses on the response of stresses near the crack tip, while the latter considers the global energy balance during crack growth. It will be shown in later chapters that these two criteria are equivalent only in elastic media.

It is apparent that the criterion Eq. (3.117) is established based on the assumption of linear elasticity with which the inverse square root singular stress field exists and the stress intensity factor is well defined. In other words, the complex fracture processes and the nonlinear deformations occurring around the crack tip are ignored. The validity of the criterion thus requires that the fracture process and nonlinear deformation zones are sufficiently small so that they are well contained inside the *K*-dominance zone around the crack tip, as shown in Figure 3.15. Thus, the value of K_c at the onset of unstable fracture taking place in the fracture process zone can be effectively employed to signify this event.

Theoretically, K_c in the criterion Eq. (3.117) should be a material constant and can be measured using standard specimens. However, K_c of a material may depend on the geometry of the test specimen and the loading condition as well. The variation of the *K*-dominance zone size is one of the causes for the variation of fracture toughness data [3-8]. Another main reason for the nonunique toughness is because the nonlinear deformation and fracture mode may depend on the geometry and loading condition. For instance, it is well known that K_c is highly sensitive to the thickness of the

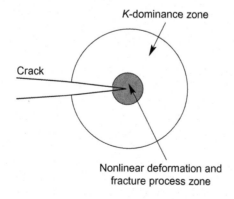

FIGURE 3.15

K-dominance zone around a crack tip.

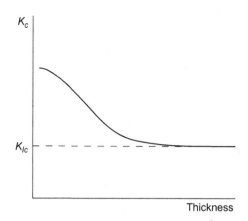

FIGURE 3.16

Thickness dependence of K_c, the critical stress intensity factor.

Table 3.1 Fracture Toughness for Typical Metals		
Materials	K_{Ic} (ksi $\sqrt{in}$)	K_{Ic} (MPa $\sqrt{m}$)
Cast iron	30	33
Low carbon steel	70	77
Stainless steel	200	220
Aluminum alloy (2024-T3)	30	33
Aluminum alloy (7075-T6)	25	28
Titanium alloy (Ti-6Al-4V)	50	55
Nickel alloy (Inconel 600)	100	110

test specimen. This thickness dependence of K_c results from the plastic deformation around the crack tip. Thinner specimens permit greater plastic deformation and thus yield higher values of K_c.

In general, the measured critical stress intensity factor K_c for a metal becomes a constant (K_{Ic}) only when the thickness of the specimen exceeds a certain value and the stress field near the crack assumes the state of plane strain, as illustrated in Figure 3.16. The notation K_{Ic} is commonly used to denote the fracture toughness obtained under the plane strain condition. The standards of measuring K_{Ic} have been established by ASTM (American Society for Testing and Materials).

Table 3.1 lists values of plane strain fracture toughness values for common metals under room temperature conditions. It can be seen that the fracture toughness ranges from about $25\,MPa\sqrt{m}$ to $250\,MPa\sqrt{m}$ for these materials. The toughness for a specific material, however, may differ from the listed data. For engineering ceramics and polymers, the fracture toughness is usually less than $5\,MPa\sqrt{m}$.

In structural integrity design, fracture mechanics is typically used to predict the fracture strength of a structural member containing a given crack(s), or to predict an allowable length of a crack in a structural member subjected to a given load(s). For example, the stress intensity factor for a through-thickness axial crack in a thin-walled cylindrical pressure vessel is given by

$$K_I = \frac{pR}{t}\sqrt{\pi a}$$

where p is the internal pressure, a the half crack length, R the mid-surface radius and t ($\ll R$) the wall thickness of the vessel. If the plastic deformation is insignificant, the failure load (burst pressure) can be calculated by substituting this stress intensity factor in Eq. (3.117) for a given crack size $2a_0$ as follows:

$$p_{cr} = \frac{tK_{Ic}}{R\sqrt{\pi a_0}}$$

On the other hand, if the pressure level p_0 is specified according to the design requirement, the maximum allowable crack size can be calculated as follows:

$$a_{allow} = \frac{1}{\pi}\left(\frac{t}{R}\right)^2\left(\frac{K_{Ic}}{p_0}\right)^2$$

References

[3-1] G.R. Irwin, Analysis of stresses and strains near the end of a crack traversing a plate, J. Appl. Mech. 24 (1957) 361–364.

[3-2] N.I. Muskhelishvili, Some Basic Problems of the Mathematical Theory of Elasticity, P. Noordhoff Ltd., Groningen, Holland, 1953.

[3-3] H.M. Westergaard, Bearing pressures and cracks, J. Appl. Mech. 6 (1939) 49–53.

[3-4] G.C. Sih, On the Westergaard method of crack analysis, Int. J. Fract. Mech. 2 (1966) 628–631.

[3-5] M.L. Williams, Stress singularities resulting from various boundary conditions in angular corners of plates in extension, J. Appl. Mech. 19 (1952) 526–528.

[3-6] M.L. Williams, On the stress distribuion at the base of a stationay crack, J. Appl. Mech. 24 (1957) 109–114.

[3-7] S.G. Larsson, A.J. Carlsson, Influence of non-singular stress terms and specimen geometry on small-scale yielding at crack tips in elastic-plastic materials, J. Mech. Phys. Solids 21 (1973) 263–277.

[3-8] C.T. Sun, H. Qian, Brittle fracture beyond stress intensity factor, J. Mech. Mater. Struct. 4 (4) (2009) 743–753.

PROBLEMS

3.1 Given the Airy stress function as

$$\phi = Ay^2$$

derive the corresponding stress components and displacement components. Identify the rigid body terms in the displacement.

3.2 Use the relations between Z_I, Z_{II}, and ψ, χ to derive the stress function

$$\phi = \mathrm{Re}\{\widehat{\widehat{Z}}_I\} + y I_m\{\widehat{Z}_I\} - y\,\mathrm{Re}\{\widehat{Z}_{II}\} + Ay^2$$

3.3 Given $\phi = \mathrm{Re}\{\widehat{\widehat{Z}}_I\} + y I_m\{\widehat{Z}_I\}$, derive the stress and displacement components. Use only the definition of the stress function,

$$\sigma_{xx} = \frac{\partial^2 \phi}{\partial y^2}, \qquad \sigma_{yy} = \frac{\partial^2 \phi}{\partial x^2}, \qquad \sigma_{xy} = -\frac{\partial^2 \phi}{\partial x \partial y}$$

the stress–strain relations, and the strain-displacement relations.

3.4 Find the stress function or Westergaard function that solves the problem of a crack of length $2a$ in an infinite plate. The crack surface is subjected to a uniform internal pressure p_0.

3.5 An infinite plate containing a crack of length $2a$ is subjected to a pair of compressive forces P as shown in Figure 3.8 (assume Q is zero in the figure). Show that the Westergaard function

$$Z_I = \frac{P}{\pi(z-b)}\sqrt{\frac{a^2-b^2}{z^2-a^2}}$$

is the solution and prove that the stress intensity factors at both crack tips are given by Eq. (3.76).

3.6 Consider a semi-infinite crack in an infinite plate opened by a pair of forces as shown in Figure 3.9. The corresponding Westergaard function is

$$Z_1 = \frac{P}{\pi(b+z)}\sqrt{\frac{b}{z}}$$

Prove that the stress intensity factor K_I is given by Eq. (3.81) and show that, for plane stress, the crack opening displacement is

$$\delta = \frac{4P}{\pi E}\ln\left|\frac{\sqrt{|x|}+\sqrt{b}}{\sqrt{|x|}-\sqrt{b}}\right|$$

3.7 An infinite plate containing a crack of length $2a$ is subjected to a uniform pressure p on the central portion of the crack faces from $x = -a_1$ to $x = a_1$ as shown in Figure 3.17. Determine the stress intensity factor using Eq. (3.79).

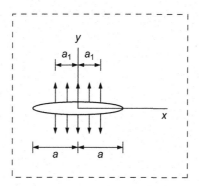

FIGURE 3.17

An infinite plate containing a crack of length $2a$ subjected to a uniform pressure p on the central portion of the crack faces.

3.8 An infinite plate containing a crack of length $2a$ is subjected to a bilinear pressure on the crack faces as shown in Figure 3.18. Determine the stress intensity factor using Eq. (3.79).

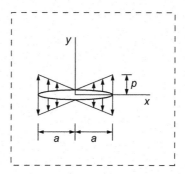

FIGURE 3.18

An infinite plate containing a crack of length $2a$ subjected to a bilinear pressure on the crack faces.

3.9 Consider an infinite plate containing an array of periodic cracks of length $2a$ subjected to biaxial tension σ_0 at infinity as shown in Figure 3.19. Show that

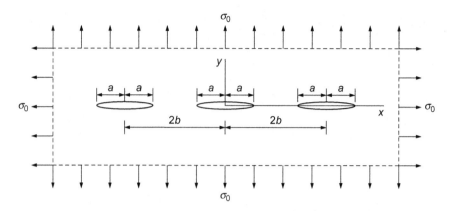

FIGURE 3.19

An infinite plate containing an array of periodic cracks of length $2a$ subjected to biaxial tension σ_0 at infinity.

the Westergaard function

$$Z_I = \sigma_0 \sin\left(\frac{\pi z}{2b}\right) \Big/ \sqrt{\sin^2\left(\frac{\pi z}{2b}\right) - \sin^2\left(\frac{\pi a}{2b}\right)}$$

is the solution and determine the stress intensity factor for the crack problem.

3.10 Consider an infinite plate with a central crack subjected to the remote tension along the y axis as shown in Figure 3.20. Find (a) σ_{yy} along the x axis (not the near crack tip solution), and (b) crack surface opening displacement u_y.

3.11 Consider Problem 3.4. Plot the degree of K-dominance along the crack line. Compare that with the case for which the plate is subjected to a remote tensile stress.

3.12 According to Williams' eigenfunction expansion for a Mode I crack, the opening stress ahead of the crack tip $(x > 0, y = 0)$ is

$$\sigma_{yy} = \frac{K_I}{\sqrt{2\pi x}} + b_0 x^{1/2} + b_1 x^{3/2}$$

Find the values of b_0 for the two loading cases in Problem 3.11.

3.13 Check the accuracy of the formula Eq. (3.82) in conjunction with Eq. (3.83) by FEA or any numerical method of your choice.

3.14 Consider Problem 3.4. Replace the internal pressure with a pair of concentrated forces applied in opposite directions at the midpoint of the crack. Assume that these forces are applied to open the crack surfaces. Compare the

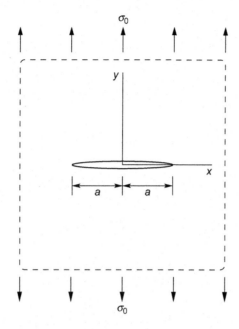

FIGURE 3.20

An infinite plate with a central crack subjected to the remote tension along the
y axis.

K-dominance zone size of this loading with that of the remote stress loading
case. If these two loading conditions are used to measure the fracture toughness of the material, which specimen would yield a higher toughness value in
terms of K_I? Explain.

Energy Release Rate

Linear elastic fracture mechanics (LEFM) can be studied by either the near-tip stress field or the energy method. In the near-tip stress field approach discussed in Chapter 3, crack growth is determined by the local stress field around the crack tip, which is characterized by the stress intensity factor. Fracture occurs when the stress intensity factor reaches its critical value, that is, the fracture toughness. It was Irwin [4-1] who introduced the concept of stress intensity factor that drew focus on the near-tip stress and displacement fields theory.

In the energy approach introduced in this chapter, the fracture behavior of a material is described by the energy variation of the cracked system during crack extension, which is characterized by the so-called energy release rate. The energy release is considered a global exercise of the cracked system. Griffith's original concept of fracture was based on the energy released during crack extension. This method was later further developed by Irwin [4-2] and Orowan [4-3]. For linear elastic materials, the energy and the stress field approaches can be considered equivalent.

4.1 THE CONCEPT OF ENERGY RELEASE RATE

Consider a two dimensional elastic body occupying an area A_0 with the boundary Γ. It is assumed that Γ consists of Γ_t and Γ_u, with Γ_t and Γ_u being the portions of the boundary with prescribed tractions and prescribed displacements, respectively. The total potential energy per unit thickness is defined as

$$\Pi = U + V \tag{4.1}$$

where

$$U = \iint_{A_0} W dA \tag{4.2}$$

is the total strain energy stored in the elastic body, and

$$V = -\int_{\Gamma_t} T_i u_i d\Gamma \tag{4.3}$$

is the potential of external forces. In Eqs. (4.2) and (4.3), T_i are the prescribed tractions on Γ_t, u_i are the corresponding displacements, and W is the strain energy density given by

$$W = W(e_{ij}) = \int_0^{e_{ij}} \sigma_{ij} de_{ij} \tag{4.4}$$

where σ_{ij} is the stress tensor and e_{ij} is the strain tensor. For linear elastic materials, W can be expressed as follows:

$$W = \frac{1}{2}\sigma_{ij}e_{ij}$$

The energy method of LEFM involves total potential energy variation of a cracked system during crack growth. The energy release rate G is defined as total potential energy decrease during unit crack extension, that is,

$$G = -\frac{d\Pi}{da} \tag{4.5}$$

where a is the crack length and da is the crack extension. If the elastic body is free from external tractions, the potential energy becomes the strain energy, that is, $\Pi = U$. The energy release rate is then equivalent to the strain energy release rate. Corresponding to the criterion Eq. (3.117) in Chapter 3, the fracture criterion in the energy approach is

$$G = G_c \tag{4.6}$$

where G_c is the critical value of G. In the Griffith theory discussed in Chapter 2, G_c is two times surface energy γ_c, which applies to perfectly britle solids. Irwin [4-2] and Orowan [4-3] extended the preceding criterion to metals experiencing small-scale yielding by lumping the surface energy and the plastic energy dissipation into G_c.

4.2 THE RELATIONS BETWEEN G AND K BY THE CRACK CLOSURE METHOD

The energy and the near-tip stress field approaches for the fracture of elastic bodies are equivalent, that is, there exists a unique relation between the energy release rate and the stress intensity factor. This relationship can be established by the so-called crack closure method.

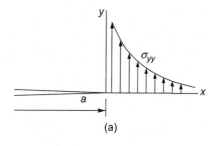

(a)

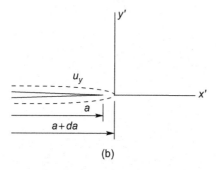

(b)

FIGURE 4.1

Virtual crack extension and crack closure method: (a) before extension of *da*, (b) after extension of *da*.

First consider a Mode I crack before and after an extension of *da* as shown in Figures 4.1(a) and (b), respectively. From (3.44) in Section 3.7, the normal stress σ_{yy} ahead of the crack tip ($\theta = 0$) (before extension) is

$$\sigma_{yy} = \frac{K_I(a)}{\sqrt{2\pi x}} \tag{4.7}$$

where $K_I(a)$ is the stress intensity factor and the origin of the coordinate system $x - y$ is at the crack tip.

After the assumed crack extension of *da*, new crack surfaces are created in $0 \le x \le da$ and the displacement of its upper face is given by (3.45), which can be written in terms of the $x' - y'$ coordinates (with the origin at the grown crack tip) as

$$u_y = \frac{\kappa + 1}{4\mu\pi} K_I \sqrt{2\pi(-x')}$$

Noting that $x' = x - da$, we rewrite this expression as

$$u_y = \frac{\kappa + 1}{4\mu\pi} K_I \sqrt{2\pi(da - x)} \tag{4.8}$$

where $K_I = K_I(a + da)$. Because da is vanishingly small, K_I in Eq. (4.8) can be taken to be equal to $K_I(a)$.

The strain energy released associated with the crack extension da can be regarded as the work done by the stress σ_{yy} in Eq. (4.7) before crack extension, which closes up the crack opening Eq. (4.8) in the region $-da \leq x' \leq 0$ after crack extension, that is, the work done by σ_{yy} in Eq. (4.7) traversing u_y in Eq. (4.8) is equal to the energy released $G_I\, da$, where G_I is the energy release rate for Mode I. We thus have

$$G_I\, da = 2 \int_0^{da} \frac{1}{2} \sigma_{yy} u_y\, dx$$

where the factor 2 on the right side accounts for the two (upper and lower) crack surfaces. Substitution of Eqs. (4.7) and (4.8) into the previous expression yields

$$G_I\, da = \frac{\kappa + 1}{4\mu\pi} K_I^2 \int_0^{da} \sqrt{\frac{1 - x/da}{x/da}}\, dx \tag{4.9}$$

With the new integral variable η defined by

$$x = da\sin^2\eta$$
$$dx = da \cdot 2\sin\eta\cos\eta\, d\eta$$

the integral in Eq. (4.9) can be evaluated as

$$\int_0^{da} \sqrt{\frac{1 - x/da}{x/da}}\, dx = 2\, da \int_0^{\pi/2} \sqrt{\frac{\cos^2\eta}{\sin^2\eta}}\, \sin\eta\cos\eta\, d\eta$$

$$= 2\, da \int_0^{\pi/2} \cos^2\eta\, d\eta$$

$$= \frac{\pi}{2}\, da$$

With this integral, we have

$$G_I\, da = \frac{\kappa + 1}{8\mu} K_I^2\, da$$

Thus,

$$G_I = \frac{\kappa + 1}{8\mu} K_I^2 \tag{4.10}$$

For plane strain, $\kappa = 3 - 4v$, and we have

$$G_I = \frac{(1-v)}{2\mu}K_I^2 = \frac{(1-v^2)}{E}K_I^2 \tag{4.11}$$

For plane stress, $\kappa = (3-v)/(1+v)$, and

$$G_I = \frac{K_I^2}{2\mu(1+v)} = \frac{K_I^2}{E} \tag{4.12}$$

For Mode II and Mode III problems, if the crack is assumed to grow in its original direction (a Mode II crack generally deflects from the original crack direction; see discussons in Chapter 5), we can obtain similar relations between G and K as

$$G_{II} = \frac{\kappa + 1}{8\mu}K_{II}^2 \tag{4.13}$$

$$G_{III} = \frac{1}{2\mu}K_{III}^2 \tag{4.14}$$

respectively.

An alternative approach for deriving the strain energy release rates is to calculate the total strain energy released during the process of crack formation from $a = 0$ to a. Consider a Mode III crack of length $2a$ in an infinite body subjected to remote shearing stress. The total strain energy may be decomposed into two parts, that is, $U_{III} = U_{IIInc} + U_{IIIc}$, where U_{IIIc} is due to the presence of the crack and U_{IIInc} is independent of crack length. According to the crack closure approach, the strain energy due to the crack is

$$U_{IIIc} = 4\int_0^a \frac{1}{2}Su_z\,dx \tag{4.15}$$

where S is the corresponding remote anti-plane shear stress. Substituting the upper crack surface displacement Eq. (3.70),

$$u_z = \frac{S}{\mu}\sqrt{a^2 - x^2}$$

in Eq. (4.15), we obtain

$$U_{IIIc} = 2\int_0^a \frac{S^2}{\mu}\sqrt{a^2 - x^2}\,dx$$

$$= 2\frac{S^2}{\mu}\left[\frac{1}{2}x\sqrt{a^2 - x^2} + \frac{1}{2}a^2\sin^{-1}\left(\frac{x}{a}\right)\right]_0^a$$

$$= \frac{\pi S^2 a^2}{2\mu} \tag{4.16}$$

The energy release rate per crack tip is

$$G_{III} = \frac{dU_{III}}{d\,(2a)} = \frac{1}{2}\frac{dU_{IIIc}}{da}$$

$$= \frac{K_{III}^2}{2\mu} \tag{4.17}$$

The relations between the strain energy release rates and the stress intensity factors allow one to calculate either parameter using the most efficient method available.

4.3 THE J-INTEGRAL

In fracture mechanics, Rice [4-4], [4-5] first derived the contour J-integral from the potential energy variation with crack extension. The J-integral theory holds for both linear and nonlinear elastic materials. The concept has been used for elastic-plastic materials obeying deformation plasticity, which will be discussed in Chapter 6.

The J-integral defined by Rice [4-4] has the following expression:

$$J = \int_{\Gamma} \left(W dy - T_i \frac{\partial u_i}{\partial x} d\Gamma \right) = \int_{\Gamma} \left(W n_1 - \sigma_{ij} \frac{\partial u_i}{\partial x} n_j \right) d\Gamma \tag{4.18}$$

where the summation convention over the dummy index i is observed and i takes 1 and 2, or x and y, Γ is an arbitrary contour beginging at the lower crack surface and ending on the upper crack surface, as shown in Figure 4.2 (Γ_1, for example), n_j are the components of the unit outward vector normal to the contour, and $T_i = \sigma_{ij}n_j$ are the tractions along the contour. It will be seen that the J-integral is the energy release rate for crack growth in an elastic body. The J-integral also represents a kind

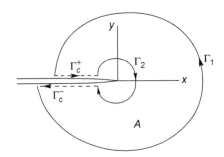

FIGURE 4.2

Integration contours in a cracked body.

of conservation property because of its path-independence. These properties of the J-integral can be proved using the following basic equations of elasticity:

$$\sigma_{ij,j} = 0 \quad \text{(equilibrium equation)} \tag{4.19}$$

$$e_{ij} = \frac{1}{2}\left(u_{i,j} + u_{j,i}\right) \quad \text{(strain-displacement relation)} \tag{4.20}$$

$$\sigma_{ij} = \frac{\partial W}{\partial e_{ij}} \quad \text{(stress–strain relation)} \tag{4.21}$$

4.3.1 *J* as Energy Release Rate

Consider a two-dimensonal cracked body with an area A_0 enclosed by the boundary Γ_0, as shown in Figure 4.3. The boundary Γ_0 consists of the outer contour Γ and the crack surfaces Γ_a. The elastic medium is subjected to the prescribed traction T_i along the boundary segment Γ_t and the prescribed displacements on the boundary segment Γ_u. The crack surfaces are along the X-axis and are assumed to be free of traction. The positive contour direction of Γ_0 is defined in that when one travels along it, the domain of interest always lies to the left of the traveler. It follows from Eqs. (4.1) to (4.3) that the potential energy Π of the cracked system per unit thickness can be written as

$$\Pi = \Pi(a) = \iint_{A_0} W dX dY - \int_{\Gamma_t} T_i u_i d\Gamma$$

where a is the crack length and (X, Y) is a stationary Cartesian coordinate system. The body forces are absent.

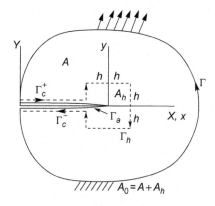

FIGURE 4.3

Contours and coordinate systems in a cracked body.

Now evaluate the energy release rate defined by

$$G = -\frac{d\Pi}{da} = -\frac{d}{da}\iint_{A_0} W dX dY + \frac{d}{da}\int_{\Gamma_t} T_i u_i d\Gamma \qquad (4.22)$$

It is known from Chapter 3 that W has a $1/r$ singularity at the crack tip (r is the distance from the tip) in LEFM. Thus, the differentiation with respect to the crack length may not be directly performed within the area integral sign in Eq. (4.22) because the differentiation with respect to a will involve the derivation with respect to r as seen in the following.

We now consider a small square A_h with the center at the crack tip as shown in Figure 4.3. The side length of the square is $2h$ and the boundary is denoted by Γ_h. The region of the cracked body excluding A_h is denoted by A, that is,

$$A_0 = A \cup A_h$$

Because no stress singularity exists in A and along Γ_t, Eq. (4.22) can be written as

$$G = -\frac{d}{da}\left[\iint_A W dX dY + \iint_{A_h} W dX dY\right] + \int_{\Gamma_t} T_i \frac{du_i}{da}d\Gamma$$

$$= -\iint_A \frac{dW}{da}dX dY + \int_{\Gamma_0} T_i \frac{du_i}{da}d\Gamma - \frac{d}{da}\iint_{A_h} W dX dY \qquad (4.23)$$

Here, the integration along Γ_t is extended to the entire boundary Γ_0 because $T_i = 0$ on the crack faces Γ_a and $du_i/da = 0$ on Γ_u. It is convenient to introduce a local coordinate system (x,y) attached at the crack tip, that is,

$$x = X - a, \quad y = Y$$

Thus,

$$\frac{d(\ )}{da} = \frac{\partial(\ )}{\partial a} - \frac{\partial(\ )}{\partial x} \qquad (4.24)$$

when the field variables are described in the local coordinate system (x,y). Using relationship Eq. (4.24), Eq. (4.23) can be written as

$$G = -\iint_A \frac{\partial W}{\partial a}dx dy + \int_{\Gamma_0} T_i \frac{\partial u_i}{\partial a}d\Gamma$$

$$+ \iint_A \frac{\partial W}{\partial x}dx dy - \int_{\Gamma_0} T_i \frac{\partial u_i}{\partial x}d\Gamma - \frac{d}{da}\iint_{A_h} W dX dY \qquad (4.25)$$

Note that on the crack surface $T_i = dy = 0$. Use of the divergence theorem thus gives

$$\iint_A \frac{\partial W}{\partial x} dA = \int_{\Gamma+\Gamma_h} W n_x d\Gamma = \int_{\Gamma+\Gamma_h} W dy \qquad (4.26)$$

and

$$\iint_A \frac{\partial W}{\partial a} dx dy = \iint_A \frac{\partial}{\partial x_j}\left(\sigma_{ij}\frac{\partial u_i}{\partial a}\right) dx dy$$

$$= \int_{\Gamma+\Gamma_h} \sigma_{ij} n_j \frac{\partial u_i}{\partial a} d\Gamma$$

$$= \int_{\Gamma+\Gamma_h} T_i \frac{\partial u_i}{\partial a} d\Gamma$$

$$= \int_{\Gamma_0+\Gamma_h} T_i \frac{\partial u_i}{\partial a} d\Gamma \qquad (4.27)$$

In deriving Eq. (4.27), the following relation has been used:

$$\frac{\partial W}{\partial a} = \frac{\partial W}{\partial e_{ij}}\frac{\partial e_{ij}}{\partial a} = \sigma_{ij}\frac{\partial}{\partial a}\left(\frac{\partial u_i}{\partial x_j}\right)$$

$$= \sigma_{ij}\frac{\partial}{\partial x_j}\left(\frac{\partial u_i}{\partial a}\right)$$

$$= \frac{\partial}{\partial x_j}\left(\sigma_{ij}\frac{\partial u_i}{\partial a}\right) - \frac{\partial u_i}{\partial a}\frac{\partial \sigma_{ij}}{\partial x_j}$$

$$= \frac{\partial}{\partial x_j}\left(\sigma_{ij}\frac{\partial u_i}{\partial a}\right) \qquad (4.28)$$

where all elasticity Eqs. (4.19), (4.20), and (4.21) have been used. Using Eqs. (4.26) and (4.27), Eq. (4.25) becomes

$$G = \int_\Gamma W dy - \int_\Gamma T_i \frac{\partial u_i}{\partial x} d\Gamma$$

$$+ \int_{\Gamma_h} W dy - \int_{\Gamma_h} T_i \frac{\partial u_i}{\partial a} d\Gamma - \frac{d}{da}\iint_{A_h} W dX dY \qquad (4.29)$$

Here again the condition that $T_i = 0$ on the crack faces Γ_a is used so that the second integral on the right side of Eq. (4.29) is reduced to the outer contour Γ.

The strain energy density function has the following universal separable form in the region near the moving crack tip:

$$W = B(a)\widetilde{W}(X - a, Y) = B(a)\widetilde{W}(x,y) \tag{4.30}$$

where $B(a)$ may depend on loading and other factors but not on the local coordinates, and $\widetilde{W}(x,y)$ is a function of local coordinates only. Now assume that A_h is so small that Eq. (4.30) holds in a region containing A_h. It has been shown in [4-6] that

$$\frac{d}{da}\iint\limits_{A_h} WdXdY = \int\limits_{\Gamma_h} Wdy + \iint\limits_{A_h} \frac{\partial W}{\partial a}dxdy$$

$$= \int\limits_{\Gamma_h} Wdy - \int\limits_{\Gamma_h} T_i\frac{\partial u_i}{\partial a}d\Gamma$$

Substituting this relation into Eq. (4.29) leads to

$$G = \int\limits_{\Gamma} \left(Wdy - T_i\frac{\partial u_i}{\partial x}d\Gamma\right)$$

The integral on the right side of this is the J-integral in Eq. (4.18). Hence, the J-integral is the energy release rate.

4.3.2 Path-Independence

In deriving the J-integral in Section 4.3.1 the integral is taken along the boundary of the edge-cracked body, excluding the crack surfaces. In fact, the integration contour Γ can be arbitrarily selected as long as it begins at the lower crack face and ends on the upper crack face, which is the original definition of Rice [4-4]. This is the path-independence of the J-integral. To prove this conservation property, choose a closed-contour Γ_T that consists of contours Γ_1, Γ_2, and Γ_c^+, Γ_c^- as shown in Figure 4.2. Clearly, there is no stress singularity in the area A enclosed by Γ_T.

Consider the contour integral along Γ_T:

$$\int\limits_{\Gamma_T} \left(Wn_x - \sigma_{ij}n_j\frac{\partial u_i}{\partial x}\right)d\Gamma$$

Applying the divergence theorem, we have

$$\int\limits_{\Gamma_T} \left(Wn_x - \sigma_{ij}n_j\frac{\partial u_i}{\partial x}\right)d\Gamma = \int\limits_{A} \left[\frac{\partial W}{\partial x} - \frac{\partial}{\partial x_j}\left(\sigma_{ij}\frac{\partial u_i}{\partial x}\right)\right]dA \tag{4.31}$$

Note from Eq. (4.28) that

$$\frac{\partial W}{\partial x} = \frac{\partial}{\partial x_j}\left(\sigma_{ij}\frac{\partial u_i}{\partial x}\right)$$

Substituting this expression into Eq. (4.31) leads to the conclusion that

$$\int_{\Gamma_T}\left(Wn_x - \sigma_{ij}n_j\frac{\partial u_i}{\partial x}\right)d\Gamma = 0 \tag{4.32}$$

Noting that the crack surfaces are traction free, Eq. (4.32) becomes

$$\circlearrowleft\int_{\Gamma_1}\left(Wdy - T_i\frac{\partial u_i}{\partial x_x}\,d\Gamma\right) + \circlearrowleft\int_{\Gamma_2}\left(Wdy - T_i\frac{\partial u_i}{\partial x}\,d\Gamma\right) = 0$$

Reversing the direction of contour Γ_2, we obtain

$$\circlearrowleft\int_{\Gamma_1}\left(Wdy - T_i\frac{\partial u_i}{\partial x}\,d\Gamma\right) = \circlearrowleft\int_{\Gamma_2}\left(Wdy - T_i\frac{\partial u_i}{\partial x}\,d\Gamma\right)$$

This implies that the J-integral

$$J = \int_{\Gamma_1}\left(Wdy - T_i\frac{\partial u_i}{\partial x}\,d\Gamma\right)$$

is path-independent.

4.3.3 Relation between J and K

Because J is the energy release rate G, the relation between G and the stress intensity factor K also holds for J and K, that is,

$$J = \frac{\kappa + 1}{8\mu}\left(K_I^2 + K_{II}^2\right) + \frac{1}{2\mu}K_{III}^2 \tag{4.33}$$

under the general mixed mode fracture cases. The preceding relationship may also be derived using Eq. (4.18) and the crack tip stress and displacement fields introduced in Chapter 3. Here the Cherepanov contour [4-7] near the crack tip is used. Since the J-integral is path-independent, we can select any path that offers the most simplicity for the integration.

Consider an integration path along the boundary of a narrow rectangle with the center at the crack tip, as shown in Figure 4.4. The length and the width of the rectangle are 2ε and 2δ, respectively. It is assumed that the rectangle is vanishingly thin

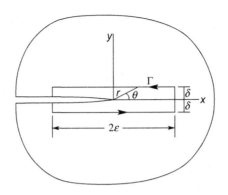

FIGURE 4.4

Cherepanov contour around a crack tip for evaluation of the J-integral.

and is within the K-dominance zone, that is, $\delta << \varepsilon$ and $\varepsilon \to 0$. Thus, the integral over the vertical lines of the contour can be neglected, and the J-integral reduces to the integral along the upper and lower horizontal lines

$$J = \lim_{\varepsilon \to 0} \int_{-\varepsilon}^{\varepsilon} \left(\sigma_{xy} \frac{\partial u_x}{\partial x} + \sigma_{yy} \frac{\partial u_y}{\partial x} + \sigma_{zy} \frac{\partial u_z}{\partial x} \right) \bigg|_{y=-\delta \to 0^-} dx$$

$$+ \lim_{\varepsilon \to 0} \int_{\varepsilon}^{-\varepsilon} \left(\sigma_{xy} \frac{\partial u_x}{\partial x} + \sigma_{yy} \frac{\partial u_y}{\partial x} + \sigma_{zy} \frac{\partial u_z}{\partial x} \right) \bigg|_{y=\delta \to 0^+} dx \qquad (4.34)$$

Note that in this integral $d\Gamma = -dx$ has been used for the upper contour. First consider the case of Mode I. Due to symmetry consideration ($\sigma_{yy}|_{y=0^-} = \sigma_{yy}|_{y=0^+}$, $u_y|_{y=0^-} = -u_y|_{y=0^+}$), Eq. (4.34) reduces to

$$J = 2 \lim_{\varepsilon \to 0} \int_{\varepsilon}^{-\varepsilon} \sigma_{yy} \frac{\partial u_y}{\partial x} \bigg|_{y=\delta \to 0^+} dx \qquad (4.35)$$

where the factor 2 accounts for the two horizontal lines.

The near-tip stress and displacement fields are

$$\sigma_{yy} = \frac{K_I}{\sqrt{2\pi r}} \cos \frac{\theta}{2} \left[1 + \sin \frac{\theta}{2} \sin \frac{3}{2}\theta \right]$$

$$u_y = \frac{K_I}{8\mu\pi} \sqrt{2\pi r} \left[(2\kappa + 1) \sin \frac{\theta}{2} - \sin \frac{3\theta}{2} \right]$$

With some mathematical manipulations, we obtain

$$\frac{\partial u_y}{\partial x} = \frac{\partial u_y}{\partial r}\frac{\partial r}{\partial x} + \frac{\partial u_y}{\partial \theta}\frac{\partial \theta}{\partial x}$$

$$= \frac{K_I}{16\mu\pi}\sqrt{\frac{2\pi}{r}}\left[-(2\kappa+2)\sin\frac{\theta}{2} + 2\sin\theta\cos\frac{3\theta}{2}\right]$$

Along the upper horizontal contour,

$$x = \delta\cot\theta$$

This relation gives

$$dx = -\frac{\delta}{\sin^2\theta}d\theta$$

Thus,

$$\sigma_{yy}\frac{\partial u_y}{\partial x}dx = \frac{K_I^2}{16\mu\pi}[(\kappa+1)\frac{\kappa+1}{2}(\cos\theta - \cos 2\theta)$$

$$- (\cos\theta + \cos 2\theta) + \frac{1}{4}(\cos 4\theta - \cos 2\theta)]d\theta$$

Substituting the previous expression into Eq. (4.35) yields the J-integral for Mode I cracks:

$$J = 2\int_0^\pi \frac{K_I^2}{16\mu\pi}(\kappa+1)d\theta = \frac{(\kappa+1)}{8\mu}K_I^2$$

Similarly, Eq. (4.33) can be proved for Mode II and Mode III cracks. This proving procedure also verifies that $J = G$.

4.3.4 Examples

Example 4.1

A semi-infinite crack in an infinite elastic strip. Consider an infinite strip containing a semi-infinite crack as shown in Figure 4.5.

The strip is stretched by a uniform vertical displacement u_0 applied along the upper and lower edges, that is,

$$u_y = \pm u_0, \quad u_x = 0 \quad \text{at } y = \pm h/2$$

Select the contour Γ for the J-integral as shown in Figure 4.5, that is, Γ consists of Γ_1, Γ_2, Γ_3, Γ_4, and Γ_5, where Γ_1 and Γ_5 are located at $x = -\infty$, Γ_3 is at $x = \infty$, and Γ_2 and Γ_4 are along the lower and upper boundary of the strip, respectively. We readily recognize the

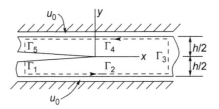

FIGURE 4.5

Integration contour in an infinite strip with a semi-infinite crack for the evaluation of J.

following properties along different segments of the contour:

$$\Gamma_1 \text{ and } \Gamma_5: \quad W = 0, \quad T_i = 0$$

$$\Gamma_2 \text{ and } \Gamma_4: \quad dy = 0, \quad \frac{\partial u_i}{\partial x} = 0$$

$$\Gamma_3: \quad \partial u_i / \partial x = 0$$

Thus, the J-integral reduces to

$$J = \int_{\Gamma_3} W_\infty \, dy = \int_{-h/2}^{h/2} W_\infty \, dy = h W_\infty$$

where W_∞ is the strain energy density at $x = \infty$ and is a constant over the height h. The uniform deformation field at $x = \infty$ is given by

$$e_{xx} = 0, \quad e_{yy} = \frac{2u_0}{h}, \quad e_{xy} = 0$$

From the stress–strain relations Eq. (3.4) for plane elasticity, we have

$$\sigma_{yy} = \frac{\mu(\kappa + 1)}{\kappa - 1} e_{yy}$$

Thus,

$$W_\infty = \frac{1}{2} \sigma_{yy} e_{yy} = \frac{2\mu(\kappa + 1)u_0^2}{(\kappa - 1)h^2}$$

and

$$J = \frac{2\mu(\kappa + 1)u_0^2}{(\kappa - 1)h}$$

Example 4.2

Double cantilever beam. Consider a double cantilever beam of unit width subjected to a symmetric pair of moments, as shown in Figure 4.6.

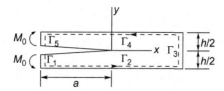

FIGURE 4.6

Integration contour in a double cantilever beam for the evaluation of *J*.

Assume that a is large as compared with h so that the simple beam theory is adequate for analysis of the split portion of the beam. The unsplit portion is also assumed to be long and, thus, the right end can be considered not loaded. Select the contour for the *J*-integral, as shown in Figure 4.6. We have

$$\text{along } \Gamma_2 \text{ and } \Gamma_4: \quad dy = 0, \quad T_i = 0$$
$$\text{along } \Gamma_3: \quad W = 0, \quad T_i = u_i = 0$$

Thus, the *J*-integral reduces to

$$J = \int_{\Gamma_1 + \Gamma_5} \left(W dy - T_i \frac{\partial u_i}{\partial x} \, d\Gamma \right)$$

$$= \int_{h/2}^{-h/2} \left(W dy + T_i \frac{\partial u_i}{\partial x} \, dy \right)$$

Note that along Γ_1 and Γ_5 the components of the unit normal vector to the contour are $n_x = -1$ and $n_y = 0$. Thus,

$$J = \int_{h/2}^{-h/2} \left(W dy - \sigma_{xx} \frac{\partial u_x}{\partial x} \, dy \right)$$

Noting that $e_{xx} = \partial u_x / \partial x$, we have

$$\sigma_{xx} \frac{\partial u_x}{\partial x} = 2W$$

and

$$J = -\int_{h/2}^{-h/2} W dy = \int_{-h/2}^{h/2} W dy$$

Because of symmetry

$$J = 2 \int_{0}^{h/2} W dy$$

Shifting the origin of the y-axis to the neutral axis of the upper leg, that is, $y' = y - h/4$, the preceding integral becomes

$$J = 2 \int_{-h/4}^{h/4} W dy'$$

For the upper leg,

$$\sigma_{xx} = \frac{M_0 y'}{I}, \qquad I = \frac{(h/2)^3}{12} = \frac{h^3}{96}$$

and

$$W = \frac{1}{2} \sigma_{xx} e_{xx} = \frac{\sigma_{xx}^2}{2E} = \frac{M_0^2 y'^2}{2EI^2}$$

We finally obtain

$$J = \frac{M_0^2}{EI}$$

which can easily be shown to be the strain energy release rate by using the compliance method.

4.4 STRESS INTENSITY FACTOR CALCULATIONS USING THE FINITE ELEMENT METHOD

The finite element method is commonly used in the stress analyses of complex engineering structures. Various techniques have been introduced to calculate stress intensity factors and energy release rates. Two commonly used methods are briefly described next.

4.4.1 Direct Method

The direct method utilizes the near-tip stress and displacement fields along the x-axis and their relations with stress intensity factors, that is,

$$\sigma_{yy} = \frac{K_I}{\sqrt{2\pi x}}, \qquad x \to 0^+$$

$$u_y = \frac{K_I(\kappa + 1)}{4\mu\pi} \sqrt{-2\pi x}, \qquad x \to 0^-$$

Thus,

$$K_I = \sigma_{yy}\sqrt{2\pi x}, \quad x \to 0^+$$

$$K_I = \frac{2\mu u_y}{\kappa + 1}\sqrt{\frac{2\pi}{-x}}, \quad x \to 0^-$$

where the origin of the x-axis resides at the crack tip. By plotting $\sigma_{yy}\sqrt{2\pi x}$ ($u_y\sqrt{-2\pi/x}$) versus x ($-x$) using the data of σ_{yy} (u_y) obtained from the finite element solution, the previous expressions indicate that the stress intensity factor K_I can be obtained within some distances from the crack tip where the plot approximately follows a horizontal straight line. In general, the value obtained from the displacement solution is more accurate.

This technique usually requires very fine element meshes in the vicinity of the crack tip in order to obtain a more accurate singular stress field. This may yield a large number of degrees of freedom and, thus, lead to more computing time. An alternative is to use special singular finite elements to surround the crack tip. Common to these singular finite elements is that they contain the proper $r^{-1/2}$ stress singularity terms in the displacement functions, which allow one to use relatively coarse meshes in the crack tip region.

4.4.2 Modified Crack Closure Technique

A popular and efficient method is to use regular finite elements to simulate the crack closure integral presented in Section 4.2. This procedure involves solving two problems, one before crack extension and the second after a virtual crack extension (see Figure 4.7). The energy released during crack extension is equal to the work done in closing the opened surfaces.

For the purpose of illustration, consider the finite element mesh near the crack tip as shown in Figure 4.7. Let the nodal forces at node $c(d)$ before crack extension be denoted by $F_x^{(c)}$ and $F_y^{(c)}$ for the horizontal and vertical components, respectively.

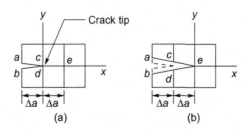

FIGURE 4.7

Virtual crack extension simulated by finite elements.

After crack extension, nodes c and d separate and their displacements are given by

$$u_x^{(c)} = \text{horizontal displacement at node } c$$

$$u_y^{(c)} = \text{vertical displacement at node } c$$

$$u_x^{(d)} = \text{horizontal displacement at node } d$$

$$u_y^{(d)} = \text{vertical displacement at node } d$$

The work done in the crack closure process can be divided into Mode I and Mode II corresponding to the opening and sliding displacements, respectively. We thus have

$$G_I = \frac{1}{2\Delta a} F_y^{(c)} \left(u_y^{(c)} - u_y^{(d)} \right)$$

$$G_{II} = \frac{1}{2\Delta a} F_x^{(c)} \left(u_x^{(c)} - u_x^{(d)} \right)$$

In this method, the virtual crack extension Δa should be small. From experience, $\Delta a/a \leq 0.05$ should be adequate.

Since $\Delta a/a$ is usually very small, the nodal displacements at nodes c and d after virtual crack extension can be approximated by the nodal displacements at nodes a and b, respectively, before virtual crack extension, that is,

$$G_I = \frac{1}{2\Delta a} F_y^{(c)} \left(u_y^{(a)} - u_y^{(b)} \right)$$

$$G_{II} = \frac{1}{2\Delta a} F_x^{(c)} \left(u_x^{(a)} - u_x^{(b)} \right)$$

Thus, there is no need to solve two problems. This is the so-called modified crack closure (MCC) method (Rybicki and Kanninen [4-8]). It should be noted that, in using the MCC method, the finite element mesh near the crack tip must be uniform. Once the energy release rates G_I and G_{II} are known, the stress intensity factors K_I and K_{II} can be calculated from Eqs. (4.10) and (4.13), respectively.

4.5 THREE-DIMENSIONAL FIELD NEAR CRACK FRONT

Most LEFM problems have been simplified and modeled as two-dimensional (2-D), though three-dimensional (3-D) effects are often acknowledged. 2-D plane strain and plane stress states are often assumed to approximate thin and thick bodies, respectively. Moreover, the plane stress formulation is not exact elasticity formulation since it does not satisfy all compatibility conditions. The error resulting from this can be significant where stress gradients are large. In view of the stress singularity at the crack tip in linearly elastic solids, the solutions of 2-D crack problems need to be reexamined from the 3-D point of view. Closed-form solutions, however,

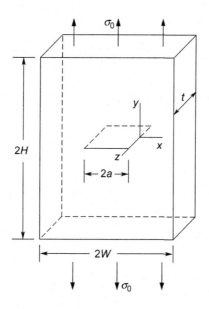

FIGURE 4.8

A through-thickness crack in an elastic plate subjected to uniform tensile stress σ_0.

are generally not available for 3-D crack problems, and finite elements and other numerical methods are usually used.

Consider a flat plate containing a center crack subjected to uniform tension as shown in Figure 4.8. Numerical solutions of the problem have been obtained in [4-9]–[4-13]. This section introduces the numerical results of Kwon and Sun [4-13] in which the dimensions of the cracked plate were taken as $W = 2a$, $H = 4a$, and $a = 12.7$ cm. The MCC technique was used to compute energy release rate and stress intensity factors.

4.5.1 Distribution of Stress Intensity Factor over Thickness

Figure 4.9 shows the opening stress σ_{yy} (normalized by $K_0/\sqrt{2\pi x}$ where $K_0 = \sigma_0\sqrt{\pi a}$) near the crack front at different locations in the thickness direction. It is evident that, near the mid-plane of the plate ($z/t \to 0$), $\sigma_{yy}\sqrt{2\pi x}$ appears to be a constant implying that the normal stress has inverse square root singularity. However, the curves of $\sigma_{yy}\sqrt{2\pi x}$ near the plate free surface ($z/t \to 0.5$) are seen to drop and then go upward as they approach the crack front. This upswing trend is more severe near the plate free surface. This is why many researchers could not obtain the nearly zero stress intensity factor at the free surface even though very fine meshes were used in the finite element analysis.

It was shown by Su and Sun [4-14] that the stress intensity factor must drop to zero at the plate surface, but this is difficult to obtain by the finite element analysis. If more

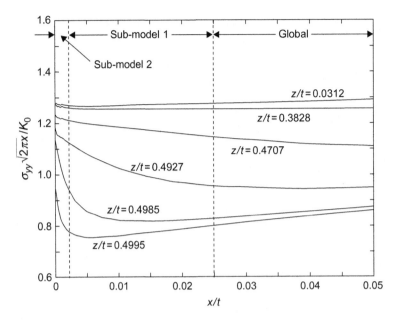

FIGURE 4.9

Normalized stress versus x/t at different z locations (adapted from Kwon and Sun [4-13]).

refined meshes in the z-direction are used, this upswing characteristic would become even more pronounced since there is another so-called (weaker) vertex singularity, which exists at the intersection between the crack front and the plate surface [4-15].

For small values of θ, Benthem [4-15] showed that the crack tip opening stress in the crack plane is expressed in the form

$$\sigma_{yy} = \frac{V}{\sqrt{2\pi\theta}}\rho^{-\lambda}$$

where ρ is the distance from the corner point, θ is the angle measured from the z-axis to the ρ direction in the $x-z$ plane, and V is the vertex stress intensity factor. For Poisson's ratio $v = 0.3$, the value of λ is around 0.45. Thus, besides the inverse square root stress singularity along the crack front, there is another stress singularity of lesser degree at the intersection point of the crack front and the plate surface (or the corner point).

Figure 4.10 shows the 3-D strain energy release rate obtained by the 3-D MCC method. 2-D plane stress and plane strain energy release rates that are obtained by the MCC are also plotted. All strain energy release rates are nondimensionalized by the 2-D plane strain energy release rate G_0:

$$G_0 = \frac{1-v^2}{E}K_0^2 = \frac{1-v^2}{E}\sigma_0^2(\pi a)$$

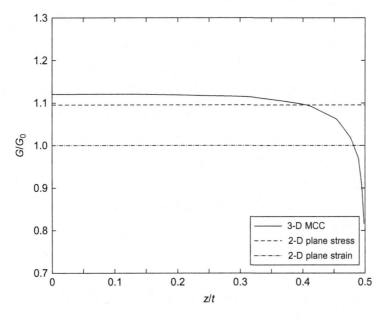

FIGURE 4.10

3-D energy release rate along the crack front and the 2-D counterparts (adapted from Kwon and Sun [4-13]).

It is shown that the strain energy release rate of the 3-D solution is closer to that of the 2-D plane stress than 2-D plane strain even though the state of the crack front is in a state of plane strain.

The stress intensity factor was obtained from the energy release rate with the plane strain conversion relation,

$$G = \frac{(1 - \nu^2)K^2}{E} \tag{4.36}$$

because the state of deformation is mostly plane strain along the crack front. Figure 4.11 shows the distribution of K over the plate thickness obtained using the 3-D energy release rate for the case $t = a$ in conjunction with the $G - K$ relation. The result obtained with the stress method is also presented in the figure. It is noted that the 3-D stress intensity factor K_{3D} at the mid-plane is larger than the 2-D stress intensity factor K_{2D}. Figure 4.12 presents the distribution of these stress intensity factors for various plate thicknesses. It is noted that, except for plates with very large thicknesses, the value of K_{2D} is quite different from that of K_{3D} at the mid-plane.

A simple technique without 3-D calculations was proposed by Kwon and Sun [4-13] for evaluating approximate values of K_{3D} at the mid-plane of the plate. The 3-D energy release rate is approximated by the 2-D plane stress (rather than plane strain) counterpart for $t \leq a$. Then this energy release rate is converted to the stress

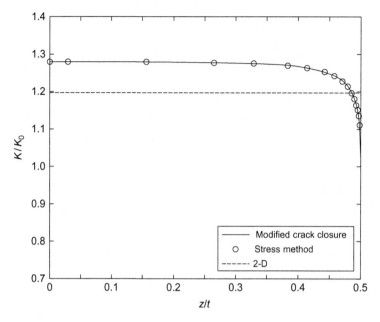

FIGURE 4.11

3-D stress intensity factor along the crack front and the 2-D counterpart (adapted from Kwon and Sun [4-13]).

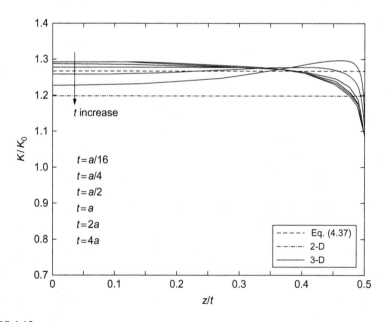

FIGURE 4.12

Stress intensity factor along the crack front for various plate thicknesses (adapted from Kwon and Sun [4-13]).

intensity factor by the plane strain $G - K$ conversion relation Eq. (4.36). With this procedure the relation between K_{3D} and K_{2D} at the mid-plane can be expressed in the form

$$\frac{K_{3D}}{K_{2D}} = \sqrt{\frac{1}{1 - \nu^2}} \tag{4.37}$$

Figure 4.12 shows the prediction of K_{3D} from K_{2D} using relation Eq. (4.37). It is evident that this method is accurate except for locations near the plate surface. Relation Eq. (4.37) could be applied not only to the center-cracked plate but also to other types of specimens such as the compact tension, three point bending, and single-edge crack specimens when $t \leq a$ [4-16].

4.5.2 Plane Strain Zone at the Crack Front

The stress field near the crack front is 3-D. However, the deformation field very close to the crack front approaches a state of plane strain, while away from the crack front it approaches a state of plane stress. Between these two states, the deformation is 3-D. The 3-D region in the cracked plate is often characterized using the parameter called

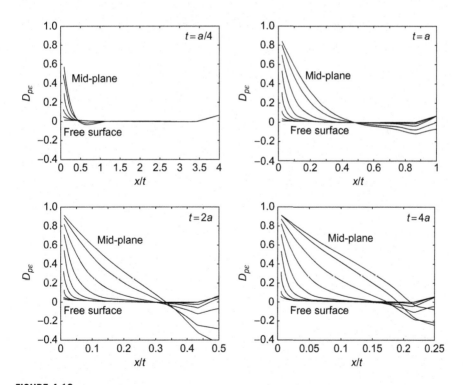

FIGURE 4.13

Degree of plane strain near the crack front for various plate thicknesses (adapted from Kwon and Sun [4-13]).

"degree of plane strain," which is defined as

$$D_{p\epsilon} = \frac{\sigma_{zz}}{\nu(\sigma_{xx} + \sigma_{yy})}$$

The parameter $D_{p\epsilon}$ is zero where the stress state is plane stress, and is unity where the stress state is plane strain. The size of the plane strain zone depends on the plate thickness.

The degrees of plane strain for different plate thicknesses are shown in Figure 4.13. In the figure, the degree of plane strain curves at different locations in the thickness direction are plotted along the crack plane. It is evident that the plane strain zone near the mid-plane of the plate is much greater than that near the plate surface. The result indicates that, for $t \leq a$, the plane strain zone size near the mid-plane of the plate is about half the plate thickness. Thus, as plate thickness decreases, the absolute size of the plane strain zone also decreases and approaches zero as $t \rightarrow 0$.

The result in Figure 4.12 clearly indicates that the stress intensity factor at the mid-plane approaches K_{2D} as the plate thickness increases beyond, say, $t = 4a$. In other words, the 2-D plane strain near-tip solution is a good approximation of the field for only very thick plates.

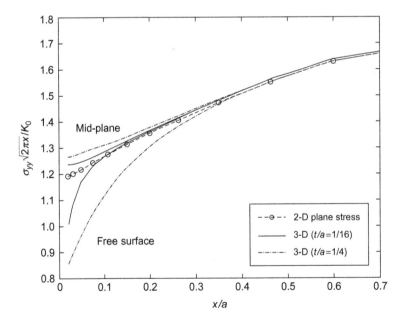

FIGURE 4.14

3-D stress field near the crack front and the 2-D plane stress counterpart (adapted from Kwon and Sun [4-13]).

On the other hand, it is shown in Figure 4.14 that 3-D stresses coincide with the plane stress solution beyond the 3-D region and that the absolute size of the 3-D region decreases as plate thickness decreases. For plates with $t \to 0$, the 3-D stress region becomes vanishingly small, and the stress intensity in the near-tip neighborhood effectively assumes the value of K_{2D}. In fact, if the 3-D region shrinks to the extent that it becomes much smaller than the fracture process zone, then the 2-D plane stress field rather than the 3-D singular field would control the fracture process.

References

[4-1] G.R. Irwin, Analysis of stresses and strains near the end of a crack traversing a plate, J. Appl. Mech. 24 (1957) 361–364.

[4-2] G.R. Irwin, Fractre dynamics, in: Fracture of Metals, American Society of Metals, Cleveland, OH, 1948, pp. 147–66.

[4-3] E. Orowan, Fundamentals of brittle fracture behavior of metals, in: Fracture and Strength of Solids, Rep. Prog. Phys. 12 (1949) 185–232.

[4-4] J.R. Rice, Mathematical analysis in the mechanics of fracture, in: H. Liebowitz (Ed.), Fracture, vol. 2, Academic Press, New York, 1968, pp. 191–311.

[4-5] J.R. Rice, A path independent integral and the approximate analysis of strain concentration by notches and cracks, J. Appl. Mech. 35 (1968) 379–386.

[4-6] Z.-H. Jin, C.T. Sun, On J-integral and potential energy variation, Int. J. Fract. 126 (2004) L19–L24.

[4-7] G.P. Cherepanov, Mechanics of Brittle Fracture, McGraw-Hill, New York, 1979, p. 240.

[4-8] E.F. Rybicki, M.F. Kanninen, A finite element calculation of stress intensity factors by a modified crack closure integral, Eng. Fract. Mech. 9 (1977) 931–938.

[4-9] D.J. Ayres, A numerical procedure for calculating stress and deformation near a slit in a three-dimensional elastic-plastic solid, Eng. Fract. Mech. 2 (1970) 87–106.

[4-10] T.A. Cruse, Three-dimensional elastic stress analysis of a fracture specimen with an edge crack, Int. J. Fract. Mech. 7 (1971) 1–15.

[4-11] T. Nakamura, D.M. Parks, Three-dimensional stress field near the crack front of a thin elastic plate, J. Appl. Mech. 55 (1988) 805–813.

[4-12] K.N. Shivakumar, I.S. Raju, Treatment of singularities in cracked bodies, Int. J. Fract. 45 (1990) 159–178.

[4-13] S.W. Kwon, C.T. Sun, Characteristics of three-dimensional stress fields in plates with a through-the-thickness crack, Int. J. Fract. 104 (2000) 291–315.

[4-14] X.M. Su, C.T. Sun, On singular stress at the crack tip of a thick plate under in-plane loading, Int. J. Fract. 82 (1996) 237–252.

[4-15] J.P. Benthem, State of stress at the vertex of a quarter-infinite crack in a half-space, Int. J. Solids Struct. 13 (1977) 479–492.

[4-16] G. Yagawa, T. Nishioka, Three-dimensional finite element analysis for through-wall crack in thick plate, Int. J. Numer. Methods Eng. 12 (1978) 1295–1310.

PROBLEMS

4.1 Use the *J*-integral and the compliance method to calculate the strain energy release rate for the double cantilever beam shown in Figure 4.15. Assume the thickness of the beam to be unity.

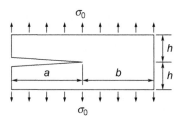

FIGURE 4.15

A double cantilever beam under uniform transverse loading.

4.2 Use plane strain or plane stress 2-D finite elements to model the split beam of Figure 4.15. Calculate the strain energy release rate using the crack closure method and compare with the results obtained in Problem 4.1. Also use the near-tip stresses obtained from the finite element analysis to extract the stress intensity factor. Compare with the results obtained in Problem 4.1 using other methods. Assume $a=b=10$ cm, $h=1$ cm, $\sigma_0=2$ MPa, $E=70$ GPa, $\nu=0.25$.

4.3 Find G_I and G_{II} of the split beam (of unit width) loaded as shown in Figure 4.16 by using the compliance method described in Chapter 2 and by the *J*-integral.

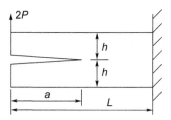

FIGURE 4.16

A split beam under a concentrated load.

4.4 Use the Mode I near-tip solutions for the stress and displacement fields to calculate the *J*-integral, and show that $G_I = K_I/E$ for plane stress problems.

4.5 Use plane stress finite elements to model the problem given in Figure 4.16, and calculate G_I and G_{II} using the MCC method and the compliance method. Consider the following two cases:

1. $a = 5$ cm, $h = 2$ cm, $L = 20$ cm, $P = 2$ kN

2. $a = 12$ cm, $h = 2$ cm, $L = 20$ cm, $P = 2$ kN

Compare the finite elements results with those obtained in Problem 4.4.

4.6 Use plane strain or plane stress 2-D finites elements to model the edge-cracked plate loaded as shown in Figure 4.17. Extract the stress intensity factor and compare the results using the following.

1. Stresses ahead of the crack tip (plot $\sqrt{2\pi x}\sigma_{yy}$ vs x curve near the crack tip).

2. Displacements of the crack surface (plot $\frac{4\mu u_y}{\kappa+1}\sqrt{\frac{\pi}{2r}}$ vs x curve near the crack tip).

3. Crack closure method.

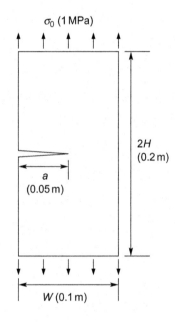

FIGURE 4.17

An edge cracked plate subjected to tension.

Mixed Mode Fracture

Prediction of crack growth generally involves two aspects, that is, when would the crack start to grow, and along what direction. Under Mode I conditions, it is well-known that the crack grows in its original direction. The fracture criterion thus can be established using either K_I in Irwin's stress intensity factor theory, or G_I in the energy release theory including the Griffith theory. Under general mixed mode conditions, experimental observations indicate that the crack will no longer grow in its original direction and K_{II} also plays a role in crack growth. Hence, appropriate fracture criteria need to be established to predict mixed mode crack extension.

As in the Mode I fracture, both energy and near-tip stress field approaches are used in mixed mode fracture studies. In the energy balance approach, a small kink from the main crack tip is analyzed to determine the energy release rate along the direction of the kink. The crack is assumed to grow along the direction that maximizes the energy release rate. This approach requires tedious mathematical analyses for a branched crack in a stress field of combined modes.

On the other hand, phenomenological models based on the stress intensity factors at the main crack tip could be established along with experimental calibration in the stress field approach. The crack growth direction is predicted directly using the existing stress intensity factor solutions, thus avoiding the complicated analyses for branched cracks. These models include the maximum tensile stress and strain energy density criteria.

5.1 A SIMPLE ELLIPTICAL MODEL

For Mode I fracture, it is known that fracture criterion can be expressed as

$$K_I = K_{Ic} \tag{5.1}$$

or equivalently

$$G_I = G_{Ic} \tag{5.2}$$

An extension of Eq. (5.1) to I/II mixed mode fracture is

$$F_{12}(K_I, K_{II}) = 0 \tag{5.3}$$

where F_{12} is a function of K_I, K_{II}, and some material constants. A simple-minded attempt to establish the functional form of F_{12} is to use the total energy release rate,

$$G = \frac{\kappa + 1}{8\mu} \left(K_I^2 + K_{II}^2 \right) \tag{5.4}$$

to gauge the critical condition for crack growth, that is, a crack extends if

$$G = G_c \tag{5.5}$$

where G_c is the critical value of G. Substituting Eq. (5.4) into Eq. (5.5) yields

$$\left(\frac{K_I}{K_c} \right)^2 + \left(\frac{K_{II}}{K_c} \right)^2 = 1 \tag{5.6}$$

where K_c is given by

$$K_c = \sqrt{\frac{8\mu G_c}{\kappa + 1}}$$

This equation represents a circle with the radius K_c in the $K_I - K_{II}$ plane. The mixed mode fracture criterion Eq. (5.6) should include Mode I and Mode II as special cases, which means $K_{Ic} = K_{IIc} = K_c$, where K_{IIc} is the crtical value of K_{II} and is in general related to K_{Ic}.

In practice, K_I and K_{II} at fracture usually do not follow the trajectory of a circle and in general $K_{Ic} \neq K_{IIc}$. The reason is that the criterion Eq. (5.6) is based on the total energy release rate formula Eq. (5.4), which holds true only when the crack grows in its original direction. Under mixed mode conditions, however, the crack does not grow in its original direction.

An extension of Eq. (5.6) to account for the disparity between K_{Ic} and K_{IIc} is to use the following functional form of K_I and K_{II}:

$$\left(\frac{K_I}{K_{Ic}} \right)^2 + \left(\frac{K_{II}}{K_{IIc}} \right)^2 = 1 \tag{5.7}$$

This equation describes an ellipse in the $K_I - K_{II}$ plane. While K_{Ic} is the fracture toughness for Mode I, K_{IIc} is determined by best matching Eq. (5.7) with the experimental data. It is noted that the crack growth direction is not discussed in the criterion Eq. (5.7).

Combined Mode I and Mode II tests are usually carried out on a tension panel with an oblique crack as shown in Figure 5.1. The in-plane sizes of the panel are much

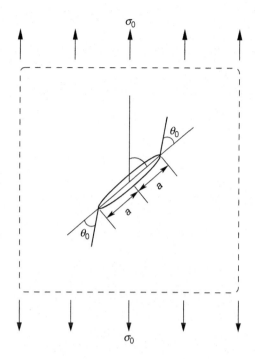

FIGURE 5.1

A tension panel with an oblique crack. The in-plane sizes of the panel are much larger than the crack length.

larger than the crack length so that the panel can be theoretically treated as an infinite one. Results of such tests on aluminum alloy sheet specimens generally confirm the applicability of Eq. (5.7) to predict mixed mode crack initiation. Figure 5.2 shows the test data for aluminum sheets (Hoskin et al. [5-1]; Pook [5-2]) and the theoretical curves from Eq. (5.7). Both plane strain and plane stress were considered.

For plane strain, the material is DTD 5050 aluminum alloy, with $K_{Ic} = 28.5$ MPa$\sqrt{\text{m}}$ and $K_{IIc} = 21.5$ MPa$\sqrt{\text{m}}$ used in Eq. (5.7). For plane stress, the material is 2024-T3 aluminum alloy and the two critical values used in the theoretical curve are 67 MPa$\sqrt{\text{m}}$ and 48.5 MPa$\sqrt{\text{m}}$, respectively. It appears from the figure that for the materials considered, K_{Ic} and K_{IIc} have an approximate correlation:

$$K_{IIc} \approx 0.75 K_{Ic}$$

Substituting this relation into Eq. (5.7) yields

$$K_I^2 + 1.78 K_{II}^2 = K_{Ic}^2$$

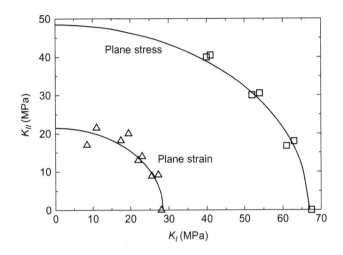

FIGURE 5.2

Experimental crack initiation data for aluminum sheets (adapted from Hoskin et al. [5-1]; Pook [5-2]) and the theoretical curves from Eq. (5.7).

5.2 MAXIMUM TENSILE STRESS CRITERION (MS-CRITERION)

Erdogan and Sih [5-3] proposed a maximum tensile stress criterion for mixed mode fracture. The criterion assumes that (1) crack extension occurs in the direction at which the circumferential stress $\sigma_{\theta\theta}$ takes the maximum with respect to θ near the crack tip, and (2) fracture takes place when $(\sigma_{\theta\theta})_{max}$ is equal to the stress that leads to Mode I fracture.

Recall the near-tip stress field:

$$
\begin{aligned}
\sigma_{xx} &= \frac{K_I}{\sqrt{2\pi r}} \cos\frac{1}{2}\theta \left[1 - \sin\frac{1}{2}\theta \sin\frac{3}{2}\theta \right] \\
&\quad - \frac{K_{II}}{\sqrt{2\pi r}} \sin\frac{1}{2}\theta \left[2 + \cos\frac{\theta}{2}\cos\frac{3}{2}\theta \right] \\
\sigma_{yy} &= \frac{K_I}{\sqrt{2\pi r}} \cos\frac{1}{2}\theta \left[1 + \sin\frac{1}{2}\theta \sin\frac{3}{2}\theta \right] \\
&\quad + \frac{K_{II}}{\sqrt{2\pi r}} \sin\frac{\theta}{2}\cos\frac{\theta}{2}\cos\frac{3}{2}\theta \\
\sigma_{xy} &= \frac{K_I}{\sqrt{2\pi r}} \sin\frac{\theta}{2}\cos\frac{\theta}{2}\cos\frac{3}{2}\theta \\
&\quad + \frac{K_{II}}{\sqrt{2\pi r}} \cos\frac{1}{2}\theta \left[1 - \sin\frac{1}{2}\theta \sin\frac{3}{2}\theta \right]
\end{aligned}
\tag{5.8}
$$

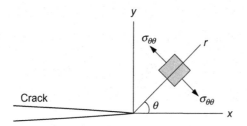

FIGURE 5.3

Coordinate systems at the crack tip and the circumferential stress $\sigma_{\theta\theta}$.

where (r,θ) are the polar coordinates centered at the crack tip as shown in Figure 5.3. Using the coordinate transformation relations of stress tensor,

$$
\begin{Bmatrix} \sigma_{xx} \\ \sigma_{yy} \\ \sigma_{xy} \end{Bmatrix} = \begin{bmatrix} \cos^2\theta & \sin^2\theta & -\sin 2\theta \\ \sin^2\theta & \cos^2\theta & \sin 2\theta \\ \frac{1}{2}\sin 2\theta & -\frac{1}{2}\sin 2\theta & \cos 2\theta \end{bmatrix} \begin{Bmatrix} \sigma_{rr} \\ \sigma_{\theta\theta} \\ \sigma_{r\theta} \end{Bmatrix}
$$

we obtain the near-tip stresses σ_{rr}, $\sigma_{\theta\theta}$, and $\sigma_{r\theta}$ in polar coordinates as follows:

$$
\begin{aligned}
\sigma_{rr} = {} & \frac{K_I}{\sqrt{2\pi r}}\left(\frac{5}{4}\cos\frac{\theta}{2} - \frac{1}{4}\cos\frac{3\theta}{2}\right) \\
& + \frac{K_{II}}{\sqrt{2\pi r}}\left(-\frac{5}{4}\sin\frac{\theta}{2} + \frac{3}{4}\sin\frac{3\theta}{2}\right) \\
\sigma_{\theta\theta} = {} & \frac{K_I}{\sqrt{2\pi r}}\left(\frac{3}{4}\cos\frac{\theta}{2} + \frac{1}{4}\cos\frac{3\theta}{2}\right) \\
& + \frac{K_{II}}{\sqrt{2\pi r}}\left(-\frac{3}{4}\sin\frac{\theta}{2} - \frac{3}{4}\sin\frac{3}{2}\theta\right) \\
\sigma_{r\theta} = {} & \frac{K_I}{\sqrt{2\pi r}}\left(\frac{1}{4}\sin\frac{\theta}{2} + \frac{1}{4}\sin\frac{3\theta}{2}\right) \\
& + \frac{K_{II}}{\sqrt{2\pi r}}\left(\frac{1}{4}\cos\frac{\theta}{2} + \frac{3}{4}\cos\frac{3}{2}\theta\right)
\end{aligned}
$$

The stresses $\sigma_{\theta\theta}$ and $\sigma_{r\theta}$ can be further simplified by using

$$
\cos 3x = 4\cos^3 x - 3\cos x
$$

$$
\sin 3x = 3\sin x - 4\sin^3 x
$$

as follows:

$$\sigma_{\theta\theta} = \frac{1}{\sqrt{2\pi r}} \cos\frac{\theta}{2} \left[K_I \cos^2\frac{\theta}{2} - \frac{3}{2} K_{II} \sin\theta \right]$$

$$\sigma_{r\theta} = \frac{1}{\sqrt{2\pi r}} \cos\frac{\theta}{2} \left[\frac{1}{2} K_I \sin\theta + \frac{1}{2} K_{II} (3\cos\theta - 1) \right]$$

(5.9)

It follows from these stress expressions that

$$\frac{\partial \sigma_{\theta\theta}}{\partial \theta} = -\frac{3}{2} \sigma_{r\theta}$$

which means that $\sigma_{\theta\theta}$ reaches its maximum $(\sigma_{\theta\theta})_{max}$ at $\theta = \theta_0$ where $\sigma_{r\theta} = 0$. The crack initiation angle θ_0 (measured from the x-axis) thus satisfies the following equation:

$$\cos\frac{\theta_0}{2} [K_I \sin\theta_0 + K_{II}(3\cos\theta_0 - 1)] = 0 \qquad (5.10)$$

or

$$\cos\frac{\theta_0}{2} = 0$$

$$K_I \sin\theta_0 + K_{II}(3\cos\theta_0 - 1) = 0$$

(5.11)

The solutions of the first equation in Eq. (5.11) are $\theta_0 = \pm\pi$, which correspond to the crack surfaces where $\sigma_{\theta\theta} = 0$. Hence, the fracture angle should be determined from the second equation in Eq. (5.11).

Now we turn to the determination of fracture load. According to the assumption of Erdogan and Sih [5-3] that fracture takes place when $(\sigma_{\theta\theta})_{max}$ satisfies

$$(\sigma_{\theta\theta})_{max} = \frac{K_{Ic}}{\sqrt{2\pi r}}$$

Then substituting the first expression in Eq. (5.9) (at $\theta = \theta_0$) into this condition yields the fracture criterion satisfied by K_I and K_{II}:

$$K_I \cos^2\frac{\theta_0}{2} - \frac{3}{2} K_{II} \sin\theta_0 = K_{Ic}/\cos\frac{\theta_0}{2} \qquad (5.12)$$

Equations (5.11) and (5.12) are the mixed mode fracture conditions in the maximum stress criterion. It is noted that only one material property, K_{Ic}, appears in the criterion.

Consider two special cases: Mode I and Mode II, respectively. For Mode I, $K_{II} = 0$, the crack extension direction is obtained from Eq. (5.11) as

$$\theta_0 = 0$$

that is, the crack extends in its original direction. For Mode II, $K_I = 0$, and the crack growth direction is now determined as

$$\theta_0 = \arccos(1/3) \approx -70.5^0$$

5.3 STRAIN ENERGY DENSITY CRITERION (S-CRITERION)

Sih [5-4] proposed a mixed mode fracture criterion based on the strain energy density concept. For a two-dimensional cracked body, the strain energy stored in an element $dV = dxdy$ of unit thickness is

$$dU = \left[\frac{1}{2\mu} \left(\frac{\kappa+1}{8} \left(\sigma_{xx}^2 + \sigma_{yy}^2 \right) - \frac{3-\kappa}{4} \sigma_{xx}\sigma_{yy} + \sigma_{xy}^2 \right) \right] dV$$

where $\kappa = 3 - 4\nu$ for plane strain and $\kappa = (3-\nu)/(1+\nu)$ for plane stress. Substituting the the crack tip stress field Eq. (5.8) for I/II mixed mode into the preceding expression, we have the strain energy density function in the form

$$\frac{dU}{dV} = W = \frac{1}{r} \left(a_{11}K_I^2 + 2a_{12}K_I K_{II} + a_{22}K_{II}^2 \right) \tag{5.13}$$

where the coefficients a_{ij} $(i,j = 1,2)$ are given by

$$a_{11} = \frac{1}{16\mu\pi} [(\kappa - \cos\theta)(1 + \cos\theta)]$$

$$a_{12} = \frac{1}{16\mu\pi} \sin\theta[2\cos\theta - (\kappa - 1)] \tag{5.14}$$

$$a_{22} = \frac{1}{16\mu\pi} [(\kappa + 1)(1 - \cos\theta) + (1 + \cos\theta)(3\cos\theta - 1)]$$

Equation (5.13) indicates that the strain energy density function has a $1/r$ singularity near the crack tip and the field intensity can be represented by the following strain energy density factor introduced by Sih [5-4]:

$$S = a_{11}K_I^2 + 2a_{12}K_I K_{II} + a_{22}K_{II}^2 \tag{5.15}$$

It is noted that the strain energy density factor S is a function of θ.

The fundamental hypotheses of the strain energy density criterion are (1) the crack will extend in the direction of minimum strain energy density (with respect to θ); and (2) crack extension occurs when the minimum strain energy density factor, $S = S_{min}$, reaches a critical value, say S_{cr}.

It is seen from Eqs. (5.13) and (5.15) that the strain energy density W and the strain energy density factor S have the same θ-dependence. Hence, the sufficient and necessary conditions for W to take the minimum value at $\theta = \theta_0$ are

$$\frac{\partial S}{\partial \theta} = 0 \quad \text{at } \theta = \theta_0 \tag{5.16}$$

$$\frac{\partial^2 S}{\partial \theta^2} > 0 \quad \text{at } \theta = \theta_0 \tag{5.17}$$

Equations (5.16) and (5.17) determine the crack extension angle θ_0. The fracture initiation condition is then characterized by

$$S(\theta_0) = S_{cr} \tag{5.18}$$

We now apply the strain energy density criterion to two special cases, that is, an infinite plate with a central crack subjected to remote tension σ (Mode I) and shear τ (Mode II), respectively.

Mode I
For the Mode I problem shown in Figure 5.4, we have

$$K_I = \sigma_0 \sqrt{\pi a}, \quad K_{II} = 0$$

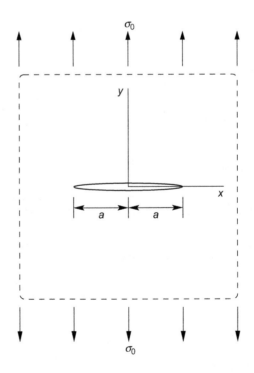

FIGURE 5.4

Crack initiation angle ($\theta_0 = 0$) in Mode I.

and

$$S = \frac{\sigma_0^2 a}{16\mu}(1+\cos\theta)(\kappa - \cos\theta)$$

The necessary condition for S to take the minimum is

$$\frac{\partial S}{\partial \theta} = 0 \Rightarrow \sin\theta\,[2\cos\theta - (\kappa - 1)] = 0$$

This equation has two solutions: $\theta_0 = 0$ and $\theta_0 = \arccos[(\kappa - 1)/2]$. Because

$$\frac{\partial^2 S}{\partial \theta^2} > 0 \quad \text{at } \theta_0 = 0$$

$$\frac{\partial^2 S}{\partial \theta^2} < 0 \quad \text{at } \theta_0 = \arccos\left(\frac{\kappa - 1}{2}\right)$$

S thus takes the minimum at $\theta_0 = 0$, that is, the crack grows in its original direction. The minimum value of S can be obtained as

$$S_{\min} = \frac{(\kappa - 1)\sigma_0^2 a}{8\mu} \quad \text{at } \theta_0 = 0°$$

From this value and the second hypothesis we get the critical applied stress:

$$\sigma_{cr} = [8\mu S_{cr}/(\kappa - 1)a]^{1/2}$$

The critical value S_{cr} can be related to the fracture toughness K_{Ic}. For plane strain $(\kappa = 3 - 4\nu)$,

$$K_{Ic}^2 = \left(\sigma_{cr}\sqrt{\pi a}\right)^2 = [4\mu\pi/(1 - 2\nu)]S_{cr}$$

Hence

$$S_{cr} = \frac{1 - 2\nu}{4\mu\pi}K_{Ic}^2$$

Mode II
For the Mode II problem as shown in Figure 5.5, we have

$$K_I = 0, \quad K_{II} = \tau_0\sqrt{\pi a}$$

$$S = \frac{\tau_0^2 a}{16\mu}[(\kappa + 1)(1 - \cos\theta) + (1 + \cos\theta)(3\cos\theta - 1)]$$

For S to take the minimum, we have

$$\frac{\partial S}{\partial \theta} = 0 \Rightarrow \sin\theta\,[(\kappa - 1) - 6\cos\theta] = 0$$

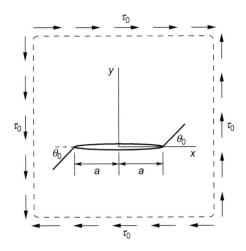

FIGURE 5.5

Crack initiation angle (θ_0) in Mode II.

Table 5.1 Crack Initiation Angle in Mode II Based on the S-Criterion					
ν	0.0	0.1	0.2	0.3	0.4	0.5
θ_0	$-70.5°$	$-75.6°$	$-79.3°$	$-83.3°$	$-87.2°$	$-90.0°$

The preceding equation has two sets of solutions:

$$\theta_0 = 0$$

$$\theta_0 = \mp \arccos\left(\frac{\kappa - 1}{6}\right)$$

It can be verified that the solution $\theta_0 = 0$ gives S_{max} and the second set of solutions gives S_{min}. The solution with a positive sign is discarded since in that direction $\sigma_{\theta\theta}$ is negative. The crack extension direction is thus given by

$$\theta_0 = -\arccos\left(\frac{\kappa - 1}{6}\right) \tag{5.19}$$

The corresponding S is

$$S_{min} = \frac{(14\kappa - 1 - \kappa^2)\tau_0^2 a}{192\mu}$$

Equation (5.19) shows that the crack extension direction depends on Poisson's ratio. Table 5.1 gives some values of θ_0 for different values of Poisson's ratio under plane strain conditions. It is noted that according to the maximum tensile stress criterion, the crack initiation angle is $\theta_0 = -70.5°$, which corresponds to a material with zero Poisson's ratio in the table.

5.4 MAXIMUM ENERGY RELEASE RATE CRITERION (ME-CRITERION)

The maximum energy release rate criterion is an extension of the Griffith fracture theory in that the crack will grow in the direction along which the maximum potential energy is released. A rigorous analysis of the energy release under mixed mode conditions requires consideration of a small kink from the original main crack tip, as shown in Figure 5.6, where α is the kink angle. Denote by k_I and k_{II} the stress intensity factors at the kinked crack tip. The energy release rate at the kink tip is thus

$$G_{kink} = \frac{\kappa + 1}{8\mu} \left(k_I^2 + k_{II}^2 \right) \tag{5.20}$$

Clearly, k_I and k_{II}, and hence G_{kink}, are functions of the kink angle α.

To study the crack growth direction, it is sufficient to consider a small kink, or more precisely, a kink with a length much smaller than the size of the K-dominance zone around the main crack tip. Under these considerations, k_I and k_{II} may be expressed in terms of K_I and K_{II} (stress intensity factors at the main crack tip), and the kink angle α as follows:

$$k_I = C_{11}(\alpha)K_I + C_{12}(\alpha)K_{II}$$
$$k_{II} = C_{21}(\alpha)K_I + C_{22}(\alpha)K_{II} \tag{5.21}$$

where C_{ij} $(i,j = 1,2)$ are functions of α. Determination of these functions generally requires analyses of a branched crack problem as shown in Figure 5.6. Using a

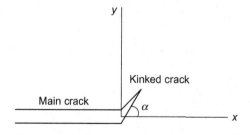

FIGURE 5.6

Crack kink in a mixed mode field.

complex stress function method, Hussain et al. [5-5] found the following expression for $C_{ij}(\alpha)$:

$$\begin{pmatrix} C_{11} & C_{12} \\ C_{21} & C_{22} \end{pmatrix} = \left(\frac{\pi - \alpha}{\pi + \alpha} \right)^{\alpha/2\pi} \left(\frac{4}{3 + \cos^2 \alpha} \right) \begin{pmatrix} \cos \alpha & \frac{3}{2} \sin \alpha \\ -\frac{1}{2} \sin \alpha & \cos \alpha \end{pmatrix} \tag{5.22}$$

It is noted that the positive angle α in Eq. (5.22) is measured clockwise from the x-axis in Figure 5.6.

Nuismer [5-6] used a continuity assumption to relate the stress intensity factors at the kinked crack tip and those at the main crack tip. A relationship of the type shown in Eq. (5.21) was obtained with the C_{ij} given by

$$C_{11} = \frac{1}{2}(1 + \cos\alpha)\cos\frac{\alpha}{2}$$

$$C_{12} = -\frac{3}{2}\sin\alpha\cos\frac{\alpha}{2}$$

$$C_{21} = \frac{1}{2}\sin\alpha\cos\frac{\alpha}{2} \tag{5.23}$$

$$C_{22} = \frac{1}{2}(3\cos\alpha - 1)\cos\frac{\alpha}{2}$$

The continuity assumption states that when the kinked crack length goes to zero, the stress field at the kinked crack tip approaches the stress field at the tip of the main crack before propagation. Later, Cotterell and Rice [5-7] used the small kink angle assumption and obtained the same results with the following different form of the expressions for C_{ij}:

$$C_{11} = \frac{1}{4}\left(3\cos\frac{\alpha}{2} + \cos\frac{3\alpha}{2} \right)$$

$$C_{12} = -\frac{3}{4}\left(\sin\frac{\alpha}{2} + \sin\frac{3\alpha}{2} \right)$$

$$C_{21} = \frac{1}{4}\left(\sin\frac{\alpha}{2} + \sin\frac{3\alpha}{2} \right) \tag{5.24}$$

$$C_{22} = \frac{1}{4}\left(\cos\frac{\alpha}{2} + 3\cos\frac{3\alpha}{2} \right)$$

They stated that these expressions are approximately valid for α up to 40 degrees.

Substituting Eq. (5.21) into Eq. (5.20) yields

$$G_{kink} = \frac{\kappa + 1}{8\mu}\left[\left(C_{11}^2 + C_{21}^2 \right)K_I^2 + \left(C_{12}^2 + C_{22}^2 \right)K_{II}^2 \right.$$

$$\left. + 2(C_{11}C_{12} + C_{21}C_{22})K_I K_{II} \right] \tag{5.25}$$

The crack growth direction, or fracture angle α_0, is thus determined by maximizing $G_{kink}(\alpha)$:

$$\frac{\partial G_{kink}(\alpha)}{\partial \alpha} = 0 \quad \text{at } \alpha = \alpha_0$$

$$\frac{\partial^2 G_{kink}(\alpha)}{\partial \alpha^2} < 0 \quad \text{at } \alpha = \alpha_0$$

(5.26)

With the expressions of C_{ij} given in Eq. (5.23) or Eq. (5.24), it can be verified that a local Mode I condition, that is, $k_{II} = 0$, prevails near the tip of the kinked crack with the direction determined by Eq. (5.26). Thus, the crack growth condition in the maximum energy release rate criterion can be expressed as

$$G_{kink}(\alpha_0) = G_{Ic} \tag{5.27}$$

Substituting Eq. (5.25) into the previous equation, we have

$$\left(C_{11}^2 + C_{21}^2\right) K_I^2 + \left(C_{12}^2 + C_{22}^2\right) K_{II}^2 + 2(C_{11}C_{12}$$
$$+ C_{21}C_{22})K_I K_{II} = \frac{8\mu}{\kappa + 1} G_{Ic} \tag{5.28}$$

Because k_{II} vanishes at the direction of the maximum energy release rate, the criterion Eq. (5.27) is equvalent to

$$k_I(\alpha_0) = C_{11}(\alpha_0)K_I + C_{12}(\alpha_0)K_{II} = K_{Ic} \tag{5.29}$$

Equation (5.29) is also called the local symmetry criterion (Cotterell and Rice [5-7]; Goldstein and Salganik [5-8]).

It can be shown that the fracture angles determined from Eq. (5.26) using the C_{ij} given in Eq. (5.23) or Eq. (5.24) are the same as that predicted from the maximum hoop stress criterion Eq. (5.11).

5.5 EXPERIMENTAL VERIFICATIONS

A tension panel with an inclined crack as shown in Figure 5.1 is a typical specimen to verify various mixed mode fracture criteria. In experiments, the crack length is usually very small compared with other specimen sizes so that the panel can be treated as an infinite one for the purpose of stress intensity factor calculations. Using the stress intensity factor solutions in Chapter 3 and the superposition method, the K_I and K_{II} for this problem can be obtained as

$$K_I = \sigma_0 \sqrt{\pi a} \sin^2 \beta$$

$$K_{II} = \sigma_0 \sqrt{\pi a} \sin \beta \cos \beta$$

(5.30)

where β is the angle between the crack and the load direction.

Substitution of Eq. (5.30) into the second equation in Eq. (5.11) yields the equation of crack growth angle by the maximum stress criterion:

$$\sin\beta \sin\theta_0 + \cos\beta(3\cos\theta_0 - 1) = 0 \qquad (5.31)$$

For the S-criterion, the strain energy density factor is

$$S = \sigma_0^2 a(a_{11}\sin^2\beta + 2a_{12}\sin\beta\cos\beta + a_{22}\cos^2\beta)\sin^2\beta$$

Hence, the crack growth angle satisfies

$$\frac{\partial S}{\partial\theta} = 0 \Rightarrow (\kappa - 1)\sin(\theta_0 - 2\beta) - 2\sin[2(\theta_0 - \beta)] - \sin 2\theta_0 = 0 \qquad (5.32)$$

Using the maximum energy release rate criterion Eqs. (5.25) and (5.26), the fracture angle $\theta_0 = \alpha_0$ is determined by

$$G_{kink}(\alpha_0) = \frac{\kappa+1}{8\mu}\left(\sigma_0^2\pi a\right)\sin^2\beta \max_{-\pi\leq\alpha\leq\pi}\left\{\left(C_{11}^2 + C_{21}^2\right)\sin^2\beta\right.$$
$$\left. + \left(C_{12}^2 + C_{22}^2\right)\cos^2\beta + (C_{11}C_{12} + C_{21}C_{22})\sin^2\beta\right\} \qquad (5.33)$$

Erdogan and Sih [5-3] performed a mixed mode fracture experiment using plexiglass sheets with oblique cracks under tension as shown in Figure 5.1. The specimen sizes are $9'' \times 18'' \times 3/16''$ and the crack length is $2''$. The angle β between the crack and the load direction ranges from $30°$ to $80°$. The initial fracture angle θ_0 at the left and right ends were measured and are denoted by $(\theta_0)_L$ and $(\theta_0)_R$, respectively. $(\theta_0)_{avg}$ is the corresponding average value. The test results are listed in Table 5.2.

Table 5.2 Measured and Calculated Values of the Fracture Angle

β		30°	40°	50°	60°	70°	80°
$(\theta_0)_L$	1	−64°	−55.5°	−50°	−40°	−29°	−17°
	2	−60°	−52°	−50°	−43.5°	−30.5°	−18°
	3	−63°	−57°	−53°	−44.5°	–	−15.5°
	4	–	−57°	−52°	−43.5°	–	–
$(\theta_0)_R$	1	−65°	−58°	−50.5°	−44°	−31.5°	−18.5°
	2	–	−53°	−52°	−40°	−31°	−17.5°
	3	−60°	−55°	−51.5°	−46°	−31.5°	−17°
	4	–	−57°	−50°	−43°	–	–
$(\theta_0)_{avg}$		−62.4°	−55.6°	−51.1°	−43.1°	−30.7°	−17.3°
MS-Criterion		−60.2°	−55.7°	−50.2°	−43.2°	−33.2°	−19.3°
S-Criterion		−63.5°	−56.7°	−49.5°	−41.5°	−31.8°	−18.5°
ME-Criterion (Eq. 5.22)		−64.6°	−60°	−54.3°	−46.6°	−36°	−20.6°

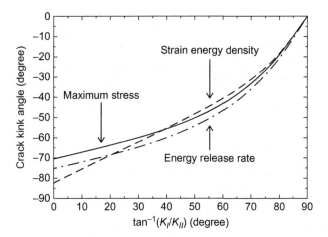

FIGURE 5.7

Fracture initiation angles predicted by the maximum tensile stress criterion, the strain energy density criterion, and the maximum energy release rate criterion (adapted from Hussain et al. [5-5]).

Also listed in the table are the theoretical predictions based on the maximum tensile stress criterion Eq. (5.31), the strain energy density criterion Eq. (5.32), and the maximum energy release rate criterion Eq. (5.33). For the maximum energy release rate criterion, only the C_{ij} given by Hussain et al. Eq. (5.22) are used since those given by Nuismer Eq. (5.23) and Cotterell and Rice Eq. (5.24) result in the same fracture angles as predicted by the maximum tensile stress criterion. It can be seen from the table that all theoretical predictions of the fracture angles are in good agreement with the experimental results.

Figure 5.7 shows the fracture angles versus the mixed mode phase angle $\tan^{-1}(K_I/K_{II})$ as predicted by the maximum tensile stress criterion, the strain energy density criterion, and the maximum energy release rate criterion. It can be observed that predictions of the strain energy density criterion deviate from those by the maximum stress and the maximum energy release rate criteria when the crack tip deformation state becomes Mode II dominated.

References

[5-1] B.C. Hoskin, D.G. Gratt, P.J. Foden, Fracture of tension panels with oblique cracks, Aeronautical Research Laboratory, Melbourne, Report SM 305, 1965.

[5-2] L.P. Pook, The effect of crack angle on fracture toughness, National Engineering Laboratory, East Kilbridge, Report NEL 449, 1970.

[5-3] F. Erdogan and G.C. Sih, On the crack extension in plates under plane loading and transverse shear, J. Basic Eng. ASME 85 (1963) 519–527.

[5-4] G.C. Sih, Strain energy density factor applied to mixed mode crack problems, Int. J. Fract. Mech. 10 (1974) 305–321.

[5-5] M.A. Hussain, S.L. Pu, J. Underwood, Strain energy release rate for a crack under combined Mode I and II, Fracture Analysis, ASTM STP 560, Am. Soc. Test. Mater. 1974, pp. 2–28.

[5-6] R.J. Nuismer, An energy release rate criterion for mixed mode fracture, Int. J. Fract. 11 (1975) 245–250.

[5-7] B. Cotterell, J.R. Rice, Slightly curved or kinked cracks, Int. J. Fract. 16 (1980) 155–169.

[5-8] R.V. Goldstein and R.L. Salganik, Brittle fracture of solids with arbitrary cracks, Int. J. Fract. 10 (1974) 507–523.

PROBLEMS

5.1 Prove that Eq. (5.10) can be obtained from $d\sigma_{\theta\theta}/d\theta = 0$.

5.2 Plot $\sigma_{\theta\theta}\sqrt{2\pi r}/K_I$ versus θ for $K_{II}/K_I = 0.5$ and 2.0, respectively. Verify that Eq. (5.11) gives the orientation at which $\sigma_{\theta\theta}$ reaches the maximum.

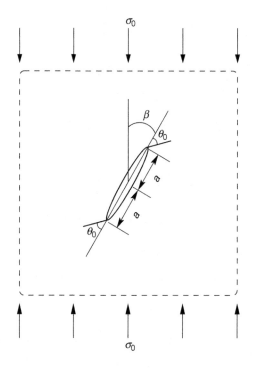

FIGURE 5.8

A compression panel with an oblique crack. The in-plane sizes of the panel are much larger than the crack length.

5.3 Plot $\sigma_{r\theta}\sqrt{2\pi r}/K_I$ versus θ for $K_{II}/K_I = 0.5$ and 2.0, respectively. Verify that $\sigma_{r\theta}$ is zero when $\sigma_{\theta\theta}$ reaches the maximum.

5.4 Show that the angle determined by $k_{II} = 0$ is along the maximum energy release rate direction.

5.5 Compare the predictions of fracture angle and failure load for a large tension panel with an oblique crack as shown in Figure 5.1. Use all methods discussed in the chapter for predictions. Assume that Poisson's ratio is 0.3 and $K_{IIc} = 1.5K_{Ic}$. Plot the results for all possible oblique angles.

5.6 Redo Problem 5.2 for a compression panel with an oblique crack as shown in Figure 5.8.

Crack Tip Plasticity

6

We have so far introduced fracture in solids undergoing linearly elastic deformations. Most engineering materials, however, exhibit various degrees of nonlinear constitutive behavior when stresses exceed a critical level. Figure 6.1 shows the typical uniaxial stress–strain curve for metals. Under loading conditions, the stress–strain relationship deviates from Hooke's law once the uniaxial stress exceeds σ_Y, which is called the yield stress. When unloading occurs at point B, the stress–strain curve follows a linear path (BB') with the slope approximately equal to the Young's modulus.

It is known from Chapter 3 that linear elastic fracture mechanics (LEFM) predicts an inverse square-root singularity at a crack tip. The stress singularity implies that the elastic stress will exceed the yield stress when approaching the crack tip, and consequently, a plastic zone develops in the crack tip region. This chapter discusses the plasticity effects around a crack tip, for example, the stress and deformation fields,

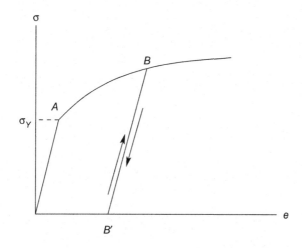

FIGURE 6.1

Uniaxial stress–strain curve for typical metals.

plastic zone size and shape, and crack tip opening displacement (CTOD). Crack initiation and growth based on these results will be discussed in Chapter 7.

6.1 YIELD CRITERIA

A yield criterion describes a condition on stresses that separates the elastic and plastic states. For the uniaxial stress state, the yield ciriterion is simply expressed as

$$\sigma = \sigma_Y$$

where σ_Y is the uniaxial yield stress; see Figure 6.1. The elastic state corresponds to $\sigma < \sigma_Y$, while the plastic deformation state prevails if $\sigma \geq \sigma_Y$. For a general three-dimensional stress state, a yield criterion is often expressed as

$$Y(\sigma_{ij}) = k$$

where Y represents a function of stresses and k is a constant. Yielding occurs if the yield criterion is satisfied. Among various yield criteria proposed, the Tresca and the von Mises criteria are most commonly used.

6.1.1 Tresca Yield Criterion

The Tresca yield criterion is based on the assumption that material yielding results from pure shear. In terms of principal stresses σ_1, σ_2, and σ_3, the maximum shear stress is among the following three quantities:

$$\tau_1 = \frac{1}{2}|\sigma_2 - \sigma_3|, \quad \tau_2 = \frac{1}{2}|\sigma_1 - \sigma_3|, \quad \tau_3 = \frac{1}{2}|\sigma_1 - \sigma_2|$$

which occur in three orthogonal planes. The Tresca criterion states that yielding occurs if the maximum shear stress reaches a critical value and can be expressed as

$$|\sigma_2 - \sigma_3| = 2\tau_{cr}, \quad \text{or}$$
$$|\sigma_1 - \sigma_3| = 2\tau_{cr}, \quad \text{or} \tag{6.1}$$
$$|\sigma_1 - \sigma_2| = 2\tau_{cr}$$

where τ_{cr} is the critical shear stress. The critical value τ_{cr} can be determined using a simple tension test in which yielding is known to occur at $\sigma_1 = \sigma_Y$ and $\sigma_2 = \sigma_3 = 0$. The last two equations in Eq. (6.1) now become

$$|\sigma_1| = 2\tau_{cr} = \sigma_Y$$

We thus have the critical shear stress

$$\tau_{cr} = \frac{\sigma_Y}{2} \tag{6.2}$$

6.1.2 **von Mises Yield Criterion**

The von Mises yield criterion assumes that material yielding is a result of distortional deformation and is independent of dilational deformation. The strain energy density W for linear elastic materials is given by

$$W = \frac{1}{2}\sigma_{ij}e_{ij}$$

Using Hooke's law, the strain energy density can be expressed in terms of the principal stresses σ_1, σ_2, and σ_3 as

$$W = \frac{1}{12\mu}\left[(\sigma_1 - \sigma_2)^2 + (\sigma_2 - \sigma_3)^2 + (\sigma_1 - \sigma_3)^2\right] + \frac{1}{18K}(\sigma_1 + \sigma_2 + \sigma_3)^2$$

where μ is the shear modulus and K is the bulk modulus. The second term associated with K in the preceding expression represents the strain energy due to volume change (dilational deformation), and the first term represents the distortional energy W_d. The von Mises yield criterion states that yielding occurs if

$$W_d = \frac{1}{12\mu}\left[(\sigma_1 - \sigma_2)^2 + (\sigma_2 - \sigma_3)^2 + (\sigma_1 - \sigma_3)^2\right] = W_{cr} \tag{6.3}$$

The critical value W_{cr} can also be obtained from a uniaxial tension test with the result

$$W_{cr} = \frac{1}{6\mu}\sigma_Y^2$$

With this determined W_{cr}, Eq. (6.3) can be rewritten as

$$\frac{1}{2}\left[(\sigma_1 - \sigma_2)^2 + (\sigma_2 - \sigma_3)^2 + (\sigma_1 - \sigma_3)^2\right] = \sigma_Y^2 \tag{6.4}$$

The von Mises criterion can also be expressed in terms of stress components in an arbitrary Cartesian coordinate system (x, y, z):

$$\frac{1}{2}\left[(\sigma_{xx} - \sigma_{yy})^2 + (\sigma_{yy} - \sigma_{zz})^2 + (\sigma_{zz} - \sigma_{xx})^2 + 6(\sigma_{xy}^2 + \sigma_{yz}^2 + \sigma_{zx}^2)\right] = \sigma_Y^2 \tag{6.5}$$

6.2 **CONSTITUTIVE RELATIONSHIPS IN PLASTICITY**

In linear elasticity, stresses and strains are related by Hooke's law:

$$\sigma_{ij} = \lambda e_{kk}\delta_{ij} + 2\mu e_{ij} \tag{6.6}$$

where λ and μ are the Lame constants, δ_{ij} is the Kronecker delta, indices i, j, and k take values $1, 2$, and 3 (or x, y, and z), and a repeated index implies summation over the range of the index. This linear relationship between stresses and strains applies until the yield condition is met. Once the yield criterion is satisfied by stresses, the constitutive law, or the stress–strain relationship, for plasticity should be adopted. Generally speaking, there is no unique stress–strain relationship in plasticity because stresses depend not only on the current strain state, but also on the strain history (note in Figure 6.1 that loading and unloading follow different paths after yielding).

A plasticity consitutive law usually relates the current stress state to strain increments, which is called the flow (or incremental) theory of plasticity. The current strain state is obtained by integrating the strain increments over the entire deformation history. The flow theory is generally used in studies of crack growth wherein unloading occurs near the crack surfaces behind the growing tip. If the material experiences only monotonic and proportional loads, a nonlinear stress–strain relationship may exist. This is the so-called deformation (or total strain) theory, which is generally used in stationary crack problems subjected to monotonically increasing loads.

6.2.1 Flow Theory of Plasticity

A basic assumption in the classical theories of plasticity is that the hydrostatic stress state does not induce plastic deformation, or that the volumetric deformation is purely elastic. The assumption has been confirmed for metals in experiments as long as the deformation is not very severe. Consequently, the volumetric and distorsional deformations can be treated separately in plasticity. First, for linear elastic materials, the relationship between the volumetric strain e_{kk} and the hydrostatic stress σ_{kk} can be obtained from Eq. (6.6) as

$$\sigma_{kk} = 3Ke_{kk} \tag{6.7}$$

Subtracting Eq. (6.7) divided by 3 from Eq. (6.6) yields

$$\sigma'_{ij} = 2\mu e'_{ij} \tag{6.8}$$

where σ'_{ij} and e'_{ij} are the deviatoric stress and strain tensors, respectively, defined by

$$\sigma'_{ij} = \sigma_{ij} - \frac{1}{3}\sigma_{kk}\delta_{ij}, \tag{6.9}$$

$$e'_{ij} = e_{ij} - \frac{1}{3}e_{kk}\delta_{ij} \tag{6.10}$$

and σ'_{ij} and e'_{ij} have only five independent components, respectively, because $\sigma'_{ii} = 0$ and $e'_{ii} = 0$. Equations (6.7) and (6.8) are thus equivalent to Eq. (6.6).

In the flow theory of plasticity, stresses are related to strain increments. The total strain increments consist of the elastic and plastic strain parts, that is,

$$de_{ij} = de^e_{ij} + de^p_{ij} \tag{6.11}$$

According to Hooke's law, in Eqs. (6.7) and (6.8), the elastic strain increments are related to the stress increments by

$$de_{ij}^e = \frac{1}{2\mu}\left[d\sigma_{ij} - \frac{\nu}{1+\nu}\delta_{ij}d\sigma_{kk}\right] \tag{6.12}$$

The plastic strain increments are derived from a plastic potential $f(\sigma_{ij})$:

$$de_{ij}^p = d\lambda\frac{\partial f}{\partial \sigma_{ij}} \tag{6.13}$$

where $d\lambda$ is a non-negative proportionality factor. The potential $f(\sigma_{ij})$ in the previous equation is often selected as the yield function. Note that the von Mises yield criterion can be expressed in the following form:

$$J_2 - \frac{1}{3}\sigma_Y^2 = 0$$

where J_2 is the second invariant of the deviatoric stress tensor, that is,

$$J_2 = \frac{1}{2}\sigma_{ij}'\sigma_{ij}'$$
$$= \frac{1}{6}\left[(\sigma_{xx} - \sigma_{yy})^2 + (\sigma_{yy} - \sigma_{zz})^2 + (\sigma_{zz} - \sigma_{xx})^2 + 6(\sigma_{xy}^2 + \sigma_{yz}^2 + \sigma_{xz}^2)\right] \tag{6.14}$$

When using the von Mises yield function as the plastic potential, the constitutive relationship Eq. (6.13) becomes

$$de_{ij}^p = d\lambda\frac{\partial J_2}{\partial \sigma_{ij}} = d\lambda\sigma_{ij}' \tag{6.15}$$

which is also called the J_2 flow theory. This constitutive law indicates that

$$\frac{de_{xx}^p}{\sigma_{xx}'} = \frac{de_{yy}^p}{\sigma_{yy}'} = \frac{de_{zz}^p}{\sigma_{zz}'} = \frac{de_{xy}^p}{\sigma_{xy}'} = \frac{de_{yz}^p}{\sigma_{yz}'} = \frac{de_{xz}^p}{\sigma_{xz}'} = d\lambda$$

It is evident from Eq. (6.15) that

$$de_{kk}^p = 0$$

Thus, the plastic deformation does not result in volume change.

To obtain the factor $d\lambda$, Eq. (6.15) is self-multiplied to result in

$$de_{ij}^p de_{ij}^p = d\lambda^2\sigma_{ij}'\sigma_{ij}'$$

We thus have

$$d\lambda = \frac{3\overline{de^p}}{2\overline{\sigma}} \tag{6.16}$$

where $\overline{\sigma}$ is the effective stress defined by

$$\overline{\sigma} = \sqrt{\frac{3}{2}\sigma'_{ij}\sigma'_{ij}} = \sqrt{3J_2} \tag{6.17}$$

and $\overline{de^p}$ is the effective plastic strain increment,

$$\overline{de^p} = \sqrt{\frac{2}{3}de^p_{ij}de^p_{ij}} = \frac{\sqrt{2}}{3}\left[(de^p_{xx} - de^p_{yy})^2 + (de^p_{yy} - de^p_{zz})^2\right.$$
$$\left. + (de^p_{zz} - de^p_{xx})^2 + 6(de^p_{xy})^2 + 6(de^p_{yz})^2 + 6(de^p_{xz})^2\right]^{1/2} \tag{6.18}$$

Substitution of Eq. (6.16) into Eq. (6.15) yields the plastic strain increments:

$$de^p_{ij} = \frac{3}{2}\frac{\overline{de^p}}{\overline{\sigma}}\sigma'_{ij} \tag{6.19}$$

By substituting the preceding equation and Eq. (6.12) into Eq. (6.11), we obtain the elastic-plastic constitutive relation in the flow theory of plasticity:

$$de_{ij} = \frac{1}{2\mu}\left[d\sigma_{ij} - \frac{\nu}{1+\nu}\delta_{ij}d\sigma_{kk}\right] + \frac{3}{2}\frac{\overline{de^p}}{\overline{\sigma}}\sigma'_{ij} \tag{6.20}$$

This equation gives the strain increments at the current loading state. To obtain the total plastic strains, one needs to integrate Eq. (6.20) over the entire loading history. The effective plastic strain increment in Eq. (6.20) may be expressed as

$$\overline{de^p} = H'd\overline{\sigma}$$

where H' is determined from the uniaxial (x-direction) stress–strain curve for which

$$\overline{\sigma} = \sigma_{xx}, \quad \overline{de^p} = de^p_x$$

It should be noted that the effective stress $\overline{\sigma}$ and effective plastic strain increment $\overline{de^p}$ cannot be independently defined. If $\overline{\sigma}$ is defined first, then the corresponding expression of $\overline{de^p}$ must be derived from the consistency in expressing the plastic work increment, that is,

$$dW^p = \sigma_{ij}de^p_{ij} = \sigma'_{ij}de^p_{ij} = \overline{\sigma}\,\overline{de^p}$$

6.2.2 Deformation Theory of Plasticity

The flow theory Eq. (6.20) applies to materials that undergo plastic deformation under general loading conditions. If the external load produces stress components that increase proportionally, Eq. (6.20) may be simplified to yield a history-independent constitutive law for loading, or the deformation theory, that relates the current state of stress to the current state of strain. Such a loading is referred to as a proportional loading.

Under proportional loading conditions, stress components vary proportionally, that is,

$$\sigma_{ij} = \tau \sigma_{ij}^0 \tag{6.21}$$

where τ is a monotonically varying parameter representing loading history and σ_{ij}^0 are independent of the loading history. Substituting these stresses into Eq. (6.15) yields the plastic strain increments:

$$de_{ij}^p = d\lambda \tau \sigma_{ij}^{0\prime}$$

Integrating this equation (over the loading history) leads to

$$e_{ij}^p = \phi \sigma_{ij}^\prime \tag{6.22}$$

where $\sigma_{ij}^\prime = t\sigma_{ij}^{0\prime}$ and

$$\phi = \frac{1}{t} \int \tau \, d\lambda$$

in which t represents the current loading state. Eq. (6.22) is actually Hencky's assumption on the relationship between plastic strains and stresses.

Self-multiplication of Eq. (6.22) yields

$$e_{ij}^p e_{ij}^p = \phi^2 \sigma_{ij}^\prime \sigma_{ij}^\prime$$

We thus obtain the factor ϕ as

$$\phi = \frac{3\overline{e^p}}{2\overline{\sigma}} \tag{6.23}$$

where

$$\overline{e^p} = \sqrt{\frac{2}{3} e_{ij}^p e_{ij}^p} \tag{6.24}$$

is the effective plastic strain. Note that under proportional deformation conditions,

$$\overline{de^p} = d\overline{e^p}$$

where $\overline{de^p}$ is given in Eq. (6.18). Substituting Eq. (6.23) into Eq. (6.22) yields

$$e_{ij}^p = \frac{3\overline{e^p}}{2\overline{\sigma}} \sigma_{ij}^\prime \tag{6.25}$$

The final stress–strain relationship in Hencky's deformation theory of plasticity can be written as

$$e'_{ij} = \left(\frac{1}{2\mu} + \frac{3}{2} \frac{\overline{e^P}}{\overline{\sigma}} \right) \sigma'_{ij} \tag{6.26}$$

$$e_{kk} = \frac{1}{3K} \sigma_{kk} \tag{6.27}$$

In Eq. (6.26), the effective plastic strain $\overline{e^P}$ can be related to the effective stress $\overline{\sigma}$ using a single curve assumption according to which the $\overline{e^P}$ versus $\overline{\sigma}$ relation is the same as that in a simple tension test. For the Ramberg-Osgood model, for example,

$$\frac{\overline{e^P}}{e_Y} = \alpha \left(\frac{\overline{\sigma}}{\sigma_Y} \right)^n$$

where α is a constant, $n > 1$ is the hardening exponent, and $e_Y = \sigma_Y/E$ is the yield strain.

6.3 IRWIN'S MODEL FOR MODE I FRACTURE

Irwin [6-1] proposed a stress relaxation model to estimate the plastic zone size ahead of the crack tip. He assumed that the stress distribution outside the plastic zone follows the elastic K-field associated with an adjusted crack length. Within the plastic zone, the stress equals the yield strength of the material. Irwin [6-1] further proposed to use an adjusted stress intensity factor as the fracture driving force for a crack in the presence of small-scale plastic deformation. The plastic deformation effect is taken into account using an effective crack length equal to the physical crack length plus half of the plastic zone size ahead of the crack tip.

6.3.1 Plastic Zone Size

Consider an elastic-perfectly plastic solid with a Mode I crack. To determine the plastic zone size along the crack line (x-axis), we first write down the stresses along the x-axis according to the elastic K-field (see Chapter 3):

$$\sigma_{xx} = \frac{K_I}{\sqrt{2\pi x}} = \sigma_1$$

$$\sigma_{yy} = \frac{K_I}{\sqrt{2\pi x}} = \sigma_2$$

$$\sigma_{zz} = \sigma_3 = \begin{cases} 0, & \text{plane stress} \\ 2\nu \frac{K_I}{\sqrt{2\pi x}}, & \text{plane strain} \end{cases} \tag{6.28}$$

The von Mises yield condition Eq. (6.4) thus becomes on the crack line

$$\sigma_{yy} = \sigma_Y^*$$
(6.29)

where σ_Y^* is given by

$$\sigma_Y^* = \begin{cases} \sigma_Y, & \text{plane stress} \\ \frac{1}{1-2v}\sigma_Y, & \text{plane strain} \end{cases}$$
(6.30)

The condition here indicates that inside the plastic zone and along the x-axis, σ_{yy} must be a constant (σ_Y in plane stress and $\sigma_Y/(1-2v)$ in plane strain). If the stress outside the plastic zone were not affected by the yielding, that is, the stress follows Eq. (6.28), the stress distribution would not be able to balance the external load if the upper half of the cracked body is taken as a free body. Hence, the stress must be redistributed outside the plastic zone due to yielding.

A simple method to resolve the problem was suggested by Irwin in which the stress σ_{yy} along the crack line still remains at σ_Y^* within the plastic zone. Outside the plastic zone, σ_{yy} is assumed to follow the K-field Eq. (6.28), but with a fictitious crack tip located at a distance of η from the physical crack tip (Figure 6.2). The fictitious crack tip position must be chosen so that the lost resultant force (represented by area A in Figure 6.2) is equal to the resultant force represented by area B. The effect of the

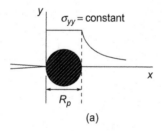

(a)

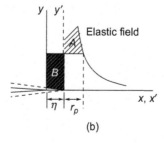

(b)

FIGURE 6.2

Irwin's model for estimating plastic zone size. (a) crack tip plastic zone, (b) fictitious crack tip at $x = \eta(x' = 0)$.

elastic-plastic stress field around a crack of length a is thus equivalent to that of an elastic K-field around a crack of length $a + \eta$.

To determine η, we use the condition area A = area B, that is,

$$\int_0^{r_p} \sigma_{yy} dx - \sigma_Y^* r_p = \eta \sigma_Y^* \qquad (6.31)$$

where r_p is obtained by substituting Eq. (6.28) into Eq. (6.29):

$$r_p = \frac{1}{2\pi} \left(\frac{K_I}{\sigma_Y^*} \right)^2 \qquad (6.32)$$

Integrating Eq. (6.31), we obtain

$$\frac{2K_I}{\sqrt{2\pi}} \sqrt{r_p} - \sigma_Y^* r_p = \eta \sigma_Y^*$$

Hence

$$\eta = r_p$$

Thus, the plastic zone size is given by

$$R_p = r_p + \eta = \frac{1}{\pi} \left(\frac{K_I}{\sigma_Y^*} \right)^2 \qquad (6.33)$$

For plane stress, $\sigma_Y^* = \sigma_Y$ and

$$r_p = \frac{K_I^2}{2\pi \sigma_Y^2} \qquad (6.34)$$

For plane strain, we have

$$r_p = (1 - 2v)^2 \frac{K_I^2}{2\pi \sigma_Y^2} \qquad (6.35)$$

Irwin suggested using

$$r_p = \frac{K_I^2}{6\pi \sigma_Y^2} \qquad (6.36)$$

for plane strain, which corresponds to $v \approx 0.2$ in Eq. (6.35).

6.3.2 **Effective Crack Length and Adjusted Stress Intensity Factor**

Irwin's model assumes that the small-scale yielding crack initiation still follows LEFM criterion with an adjusted stress intensity factor due to crack tip plasticity. The adjusted stress intensity factor is obtained using the same formula as in LEFM, but with an effective (half) crack length equal to the physical (half) crack length plus r_p, the half size of the plastic zone, ahead of each crack tip.

Two examples are given here to illustrate the concept. First consider an infinite plate with a crack of length $2a$ subjected to remote stress σ. We know the following stress intensity factor from Chapter 3:

$$\sigma \sqrt{\pi a}$$

Irwin's adjusted stress intensity is thus given by

$$K_I = \sigma \sqrt{\pi a_{\text{eff}}} \qquad (6.37)$$

where a_{eff} is half the effective crack length defined by

$$a_{\text{eff}} = a + r_p \qquad (6.38)$$

Substituting Eq. (6.32) into Eqs. (6.38) and (6.37) yields

$$K_I = \sigma \sqrt{\pi \left[a + \frac{1}{2\pi} \left(\frac{K_I}{\sigma_Y^*} \right)^2 \right]}$$

Solving for K_I in this equation leads to the adjusted stress intensity factor:

$$K_I = \sigma \sqrt{\pi a} / \sqrt{1 - \frac{1}{2} \left(\frac{\sigma}{\sigma_Y^*} \right)^2} \qquad (6.39)$$

The second example involves an edge crack of length a in a semi-infinite plate subjected to remote tension σ. The stress intensity factor expression in LEFM is

$$1.1215 \sigma \sqrt{\pi a}$$

The adjusted stress intensity factor now becomes

$$K_I = 1.1215 \sigma \sqrt{\pi \left[a + \frac{1}{2\pi} \left(\frac{K_I}{\sigma_Y^*} \right)^2 \right]}$$

Solving the previous equation, we have

$$K_I = 1.1215 \sigma \sqrt{\pi a} / \sqrt{1 - 0.6289 \left(\frac{\sigma}{\sigma_Y^*} \right)^2} \qquad (6.40)$$

Due to the simple expressions of the stress intensity factors in the two examples above, the adjusted stress intensity factors are obtained in closed forms. In general, an iteration procedure is needed to obtain the adjusted stress intensity factor. These examples show that the adjusted stress intensity factor reduces to the stress intensity factor in LEFM when $\sigma/\sigma_Y^* \ll 1$.

6.3.3 Crack Tip Opening Displacement (CTOD)

With Irwin's plastic zone model, we can define the crack opening displacement at the physical crack tip. According to Irwin's model, the physical crack tip lies on the elastic-plastic boundary and thus, the elastic solution associated with the fictitious crack tip is valid at the point. The crack opening displacement field is now given by

$$2u_y = \frac{8K_I}{E^*}\sqrt{\frac{r}{2\pi}}$$

where r is the distance from the fictitious crack tip. The CTOD, δ, is defined as

$$\delta = 2u_y\big|_{r=r_p}$$

Substituting r_p from Eq. (6.32) into the preceding equation, we obtain the CTOD in Irwin's model:

$$\delta = \frac{4}{\pi E^*}\frac{K_I^2}{\sigma_Y^*} = \frac{4G_I}{\pi \sigma_Y^*} \tag{6.41}$$

The CTOD now is proportional to the square of stress intensity factor.

6.4 THE DUGDALE MODEL

Due to the nonlinear and history-dependent nature of plasticity, it is often necessary to use simplified models to treat elastic-plastic fracture problems, for example, Irwin's model in the previous section. The Dugdale model represents another commonly used model in elastic-plastic fracture mechanics. In the Dugdale model [6-2], yielding is assumed to occur only in a narrow strip zone with zero height along the crack line. This strip yield zone is then treated as an extended part of the physical crack with the closure yield stress acting on the extended crack faces. The strip yield zone and the physical crack form the effective crack. With the Dugdale model, the elastic-plastic crack problem is reduced to an elastic one with finite stresses at the effective crack tip, which is actually a point on the elastic-plastic boundary. The Dugdale model is first introduced for small-scale yielding Mode I cracks, and is subsequently described for a centrally cracked infinite plate.

6.4.1 Small-Scale Yielding

Figure 6.3 shows a strip yield zone with a length of R_0 ahead of a crack tip. The yield zone is treated as a part of the effective crack in the Dugdale model. The applied load thus induces a stress intensity factor at the effective crack tip. Here we do not distinguish this stress intensity factor with that at the physical crack tip, that is,

$$K_I(a + R_0) = K_I(a)$$

where a is the physical crack length, because $R_0 << a$ according to small-scale yielding.

According to the Dugdale model, stresses are finite at the effective crack tip. Because the stress singularity is associated with the stress intensity factor, this finite stress condition can be met if and only if K_I due to the external loading is negated by $K_I^{(2)}$ induced by the yield stress along the strip yield zone, that is,

$$K_I + K_I^{(2)} = 0 \tag{6.42}$$

To calculate $K_I^{(2)}$, we use the known solution for a pair of concentrated forces P applied on the crack faces at a distance η from the effective crack tip as shown in Figure 6.4:

$$K_I^P = \frac{\sqrt{2}P}{\sqrt{\pi \eta}}$$

For the distributed traction σ_Y^* acting on the crack faces from $\eta = 0$ at the effective crack tip to $\eta = R_0$ at the physical crack tip, the stress intensity may be obtained by replacing P in the equation with $-\sigma_Y^* d\eta$ and integrating from $\eta = 0$ to $\eta = R_0$ as

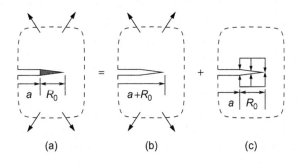

(a) (b) (c)

FIGURE 6.3

Dugdale yield zone in small-scale yielding.

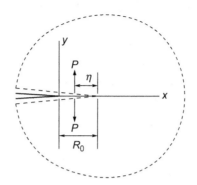

FIGURE 6.4

A pair of concentrated forces exerted on the crack faces.

follows:

$$K_I^{(2)} = \int_0^{R_0} \frac{-\sqrt{2}\sigma_Y^* d\eta}{\sqrt{\pi\eta}}$$

$$= -2\sigma_Y^*\sqrt{\frac{2R_0}{\pi}}$$

Substituting this expression into Eq. (6.42) yields

$$2\sigma_Y^*\sqrt{\frac{2R_0}{\pi}} = K_I$$

We thus obtain the length of the plastic zone,

$$R_0 = \frac{\pi}{8}\left(\frac{K_I}{\sigma_Y^*}\right)^2 \tag{6.43}$$

For plane stress, $\sigma_Y^* = \sigma_Y$,

$$R_0 \simeq 0.392\left(\frac{K_I}{\sigma_Y}\right)^2$$

This compares to the plastic zone size of Irwin's model,

$$R_p = \frac{1}{\pi}\left(\frac{K_I}{\sigma_Y}\right)^2 \simeq 0.318\left(\frac{K_I}{\sigma_Y}\right)^2$$

The opening displacement at the physical crack tip (CTOD) as a result of the applied stress intensity can be obtained from the crack tip displacement expression in Chapter 3:

$$\delta^{(1)} = \frac{8K_I}{E^*}\sqrt{\frac{r}{2\pi}}\Bigg|_{r=R_0} = \frac{2K_I^2}{E^*\sigma_Y^*}$$

The crack opening displacement due to a pair of concentrated forces P as shown in Figure 6.4 can be obtained from the Westergaard function given in Problem 3.6 as

$$\delta^{(2)} = \frac{4P}{\pi E^*}\ln\left|\frac{\sqrt{|R_0-x|}+\sqrt{\eta}}{\sqrt{|R_0-x|}-\sqrt{\eta}}\right|$$

At the physical crack tip $r = R_0$, the CTOD due to the yield stress in the plastic zone is obtained by taking $P = \sigma_Y^* d\eta$ in the previous equation and then integrating over the plastic zone, that is,

$$\delta^{(2)} = -\frac{4\sigma_Y^*}{\pi E^*}\int_0^{R_0}\ln\frac{\sqrt{R_0}+\sqrt{\eta}}{\sqrt{R_0}-\sqrt{\eta}}d\eta$$

$$= -\frac{K_I^2}{E^*\sigma_Y^*}$$

The total CTOD at the physical crack tip is thus obtained as follows:

$$\delta = \delta^{(1)} + \delta^{(2)}$$

$$= \frac{K_I^2}{E^*\sigma_Y^*} \tag{6.44}$$

This value is slightly smaller than the CTOD in Irwin's model Eq. (6.41).

6.4.2 A Crack in an Infinite Plate

Unlike Irwin's model, which is restricted to small-scale yielding, the Dugdale model also applies to large-scale yielding cases. We now consider an infinite plate with a crack of length $2a$ subjected to remote tensile stress σ, as shown in Figure 6.5. Now the effective crack has a length of $2(a+R_0)$, where R_0 is the length of the strip yield zone.

We still use the superposition method to treat the problem. The stress intensity factor at the tip of the effective crack due to the remote tension is

$$K_I = \sigma\sqrt{\pi(a+R_0)}$$

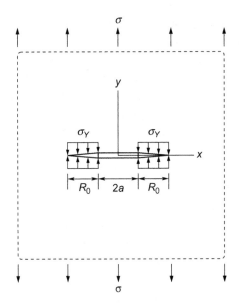

FIGURE 6.5

Dugdale yield zones ahead of crack tips in an infinite plate subjected to tension.

For the stress to remain finite at the effective crack tip, K_I should be negated by the following stress intensity factor due to the yield stress in the strip yield zone:

$$K_I^{(2)} = -2\sigma_Y^* \sqrt{\frac{a+R_0}{\pi}} \int_a^{a+R_0} \frac{d\eta}{(a+R_0)^2 - \eta^2}$$

$$= -2\sigma_Y^* \sqrt{\frac{a+R_0}{\pi}} \cos^{-1}\left(\frac{a}{a+R_0}\right)$$

The cancellation of stress singularity at the effective crack tip thus requires

$$\sigma\sqrt{\pi(a+R_0)} - 2\sigma_Y^* \sqrt{\frac{a+R_0}{\pi}} \cos^{-1}\left(\frac{a}{a+R_0}\right) = 0$$

The yield zone length is obtained from this equation:

$$\frac{R_0}{a} = \sec\left(\frac{\pi\sigma}{2\sigma_Y^*}\right) - 1 \tag{6.45}$$

Now calculate the CTOD at the physical crack tip, or the tail of the yield zone. The CTOD due to the remote tension is

$$\delta^{(1)} = \frac{4\sigma}{4E^*}\sqrt{(a+R_0)^2 - a^2}$$

This CTOD is reduced by that due to the yield stress:

$$\delta^{(2)} = -\frac{8\sigma_Y^*}{\pi E^*}\sqrt{(a+R_0)^2 - a^2}\cos^{-1}\left(\frac{a}{a+R_0}\right)$$

The total CTOD at the physical crack tip is thus obtained as

$$\delta = \delta^{(1)} + \delta^{(2)}$$
$$= \frac{8a\sigma_Y^*}{\pi E^*}\ln\left(\sec\frac{\pi\sigma}{2\sigma_Y^*}\right) \tag{6.46}$$

Equations (6.45) and (6.46) are valid for load levels up to $\sigma = \sigma_Y^*$, which corresponds to the failure of comprehensive yielding. In this case, the failure is no longer characterized by crack growth. In general, the crack will grow before σ reaches σ_Y^*. To study the yielding effect at small ratio of σ/σ_Y^*, Eqs. (6.45) and (6.46) are expanded into Taylor series as follows:

$$\frac{R_0}{a} = \frac{1}{2}\left(\frac{\pi\sigma}{2\sigma_Y^*}\right)^2 + \frac{5}{24}\left(\frac{\pi\sigma}{2\sigma_Y^*}\right)^4 + \cdots$$

$$\delta = \frac{8a\sigma_Y^*}{\pi E^*}\left[\frac{1}{2}\left(\frac{\pi\sigma}{2\sigma_Y^*}\right)^2 + \frac{1}{12}\left(\frac{\pi\sigma}{2\sigma_Y^*}\right)^4 + \cdots\right]$$

When $\sigma/\sigma_Y^* << 1$, this series can be approximated by

$$R_0 = \frac{\pi}{8}\left(\frac{K_I}{\sigma_Y^*}\right)^2 \tag{6.47}$$

$$\delta = \frac{K_I^2}{E^*\sigma_Y^*} \tag{6.48}$$

Here we have employed $K_I = \sigma\sqrt{\pi a}$ because $R_0 << a$. The preceding results about the plastic zone size and CTOD agree with the small-scale-yielding results Eqs. (6.43) and (6.44), respectively.

In general, the Dugdale model predicts crack growth based on a CTOD criterion. This will be discussed in Chapter 7. For small-scale yielding, Eqs. (6.44) and (6.48) can be rewritten as

$$K_I = \sqrt{E^*\sigma_Y^*\delta} \tag{6.49}$$

which is theoretically valid when the applied stress is much smaller than the yield stress. Burdekin and Stone [6-3] suggested using Eq. (6.49) for moderately higher load level. Thus, an LEFM approach may be combined with the Dugdale model to predict fracture.

6.5 PLASTIC ZONE SHAPE ESTIMATE ACCORDING TO THE ELASTIC SOLUTION

Knowledge of the plastic zone shape and size around a crack is useful to understanding the plasticity effect on the fracture behavior of solids. In general, complete solutions of stresses in both elastic and plastic regions are needed to determine the plastic zone shape. If the plastic zone around the crack tip is small compared to the region in which the crack tip K-fields apply, the singular elastic stress fields may be used to estimate the plastic zone shape. The estimates thus obtained, however, are the first-order approximations because the stress redistribution caused by the small plastic zone is not considered. In the following part of this section, the plastic zone shapes are estimated for Mode I cracks in both plane stress and plane strain fields. Mode II loading cases can be treated in a similar manner.

6.5.1 Principal Stresses

Use of the Tresca yield condition to determine the plastic zone requires principal stress information. We thus first determine the principal stresses based on the near-tip elastic stress fields. Under Mode I loading conditions, the K-field around the crack tip is given by

$$\sigma_{xx} = \frac{K_I}{\sqrt{2\pi r}}\left[1 - \sin\frac{\theta}{2}\sin\frac{3\theta}{2}\right]\cos\frac{\theta}{2}$$

$$\sigma_{yy} = \frac{K_I}{\sqrt{2\pi r}}\left[1 + \sin\frac{\theta}{2}\sin\frac{3\theta}{2}\right]\cos\frac{\theta}{2}$$

$$\sigma_{xy} = \frac{K_I}{\sqrt{2\pi r}}\sin\frac{\theta}{2}\cos\frac{\theta}{2}\cos\frac{3\theta}{2} \tag{6.50}$$

$$\sigma_{zz} = \begin{cases} 0, & \text{plane stress} \\ \\ \nu(\sigma_{xx} + \sigma_{yy}), & \text{plane strain} \end{cases}$$

Since $\sigma_{yz} = \sigma_{xz} = 0$, $\sigma_{zz} = \sigma_3$ is a principal stress. The other two principal stresses in the $x - y$ plane are given by

$$\left.\begin{array}{c} \sigma_1 \\ \\ \sigma_2 \end{array}\right\} = \frac{1}{2}(\sigma_{xx} + \sigma_{yy}) \pm \sqrt{\frac{1}{4}(\sigma_{xx} - \sigma_{yy})^2 + \sigma_{xy}^2}$$

Substitution of Eq. (6.50) into these expressions yields

$$\sigma_1 = \frac{K_I}{\sqrt{2\pi r}} \cos\frac{\theta}{2}\left[1 + \sin\frac{\theta}{2}\right]$$

$$\sigma_2 = \frac{K_I}{\sqrt{2\pi r}} \cos\frac{\theta}{2}\left[1 - \sin\frac{\theta}{2}\right]$$

(6.51)

Similarly, we have the principle stress σ_3,

$$\sigma_3 = \begin{cases} 0, & \text{plane stress} \\[2mm] 2\nu\frac{K_I}{\sqrt{2\pi r}}\cos\frac{\theta}{2}, & \text{plane strain} \end{cases}$$

(6.52)

6.5.2 Plane Stress Case

In the following, both Tresca and Mises yield criteria are used to estimate the crack tip plastic zone shape and size for plane stress. It can be seen that the two criteria result in different plastic zone shapes but the same plastic zone size ahead of the crack.

Plastic Zone Estimate Based on the Tresca Yield Criterion

Equations (6.51) and (6.52) indicate that $\sigma_1 \geq \sigma_2 \geq \sigma_3$ for plane stress when $\theta \geq 0$. The Tresca yield criterion Eqs. (6.1) and (6.2) now reduces to

$$\sigma_1 - \sigma_3 = \sigma_Y$$

(6.53)

Substitution of Eqs. (6.51) and (6.52) into this equation leads to

$$\frac{K_I}{\sqrt{2\pi r}}\left[1 + \sin\frac{\theta}{2}\right]\cos\frac{\theta}{2} = \sigma_Y$$

The elastic-plastic boundary is obtained by solving this equation for r as follows:

$$r = \frac{K_I^2}{2\pi\sigma_Y^2}\cos^2\frac{\theta}{2}\left[1 + \sin\frac{\theta}{2}\right]^2, \qquad \theta \geq 0$$

(6.54)

Figure 6.6 depicts the elastic-plastic boundary represented by this expression. Here the boundary for $\theta < 0$ is obtained by symmetry.

Plastic Zone Estimate Based on the von Mises Yield Criterion

Substituting Eqs. (6.51) and (6.52) into the von Mises criterion Eq. (6.4), we obtain the elastic-plastic boundary as follows:

$$r = \frac{K_I^2}{2\pi\sigma_Y^2}\cos^2\frac{\theta}{2}\left[1 + 3\sin^2\frac{\theta}{2}\right]$$

(6.55)

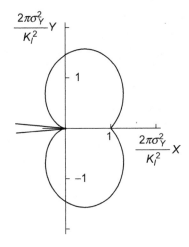

FIGURE 6.6

Plastic zone shape estimate based on the Tresca criterion in plane stress.

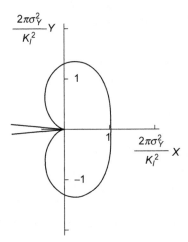

FIGURE 6.7

Plastic zone shape estimate based on the von Mises criterion in plane stress.

A plot of this expression is shown in Figure 6.7. The plastic zone size along the crack line ($\theta = 0$) is

$$R_p = r(\theta = 0) = \frac{1}{2\pi} \left(\frac{K_I}{\sigma_Y} \right)^2$$

which is half of Irwin's estimate Eq. (6.33).

6.5.3 **Plane Strain Case**

The Tresca and Mises yield conditions are again used to approximately determine the crack tip plastic zone shape and size for plane strain. We will see that the plastic zone size for plane strain is much smaller than that for the corresponding plane stress case.

Plastic Zone Estimate Based on the Tresca Yield Criterion

In the plane strain case, application of the Tresca yield criterion depends on Poisson's ratio because it influences the magnitude of σ_3 relative to σ_2. Since σ_3 can be expressed as

$$\sigma_3 = v(\sigma_1 + \sigma_2)$$

it is easy to show that when $\theta \geq 0$,

$$\sigma_1 \geq \sigma_2 \geq \sigma_3, \quad \text{if} \quad v \leq \frac{\sigma_2}{\sigma_1 + \sigma_2} \tag{6.56}$$

and

$$\sigma_1 \geq \sigma_3 \geq \sigma_2, \quad \text{if} \quad v \geq \frac{\sigma_2}{\sigma_1 + \sigma_2} \tag{6.57}$$

The corresponding functions of the elastic-plastic boundary are ($\theta \geq 0$)

$$r = \frac{K_I^2}{2\pi \sigma_Y^2} \cos^2 \frac{\theta}{2} \left(1 - 2v + \sin \frac{\theta}{2}\right)^2, \quad v \leq \frac{\sigma_2}{\sigma_1 + \sigma_2} \tag{6.58}$$

$$r = \frac{K_I^2}{2\pi \sigma_Y^2} \sin^2 \theta, \quad v \geq \frac{\sigma_2}{\sigma_1 + \sigma_2} \tag{6.59}$$

respectively. The contour of the plastic zone is to be constructed from these two functions depending on the inequalities in Eqs. (6.56) and (6.57). For $v = 1/2$, Eq. (6.57) is always satisfied. Hence, the plastic zone contour is given by Eq. (6.59) and is depicted in Figure 6.8. Again the boundary for $\theta < 0$ is obtained by symmetry.

Plastic Zone Estimates Based on the von Mises Yield Criterion

The plastic zone boundary according to the von Mises yield criterion Eq. (6.4) is obtained by substituting Eqs. (6.51) and (6.52) into Eq. (6.4):

$$r = \frac{K_I^2}{2\pi \sigma_Y^2} \cos^2 \frac{\theta}{2} \left[(1 - 2v)^2 + 3\sin^2 \frac{\theta}{2}\right] \tag{6.60}$$

This function indicates that the plastic zone shape in plane strain depends on Poisson's ratio (this is also the case with the Tresca yielding condition; see Eqs. (6.58) and (6.59)). The reason is that Poisson's ratio affects the stress triaxiality, which

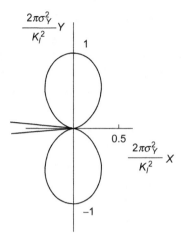

FIGURE 6.8

Plastic zone shape estimate based on the Tresca criterion in plane strain ($v = 1/2$).

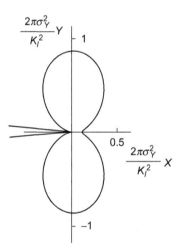

FIGURE 6.9

Plastic zone shape estimate based on the von Mises criterion in plane strain ($v = 0.3$).

restricts the plastic deformation. For example, the plastic zone size ahead of the crack tip ($\theta = 0^o$) is given by

$$R_p = r(\theta = 0) = \frac{K_I^2}{2\pi \sigma_Y^2}, \quad \text{if } v = 0$$

$$R_p = r(\theta = 0) = 0, \quad \text{if } v = 0.5$$

Figure 6.9 depicts the plastic zone shape according to Eq. (6.60) for $v = 0.3$.

A comparison of the plastic zone size along the crack line between plane stress Eq. (6.55) and plane strain Eq. (6.60) ($v = 0.3$) using the von Mises condition shows that the plane strain value of

$$R_p = r(\theta = 0) = \frac{0.16}{2\pi}\left(\frac{K_I}{\sigma_Y}\right)^2$$

is only 16% of the corresponding plane stress value,

$$R_p = r(\theta = 0) = \frac{1}{2\pi}\left(\frac{K_I}{\sigma_Y}\right)^2$$

for the same applied K_I. The smaller plastic zone size in plane strain results from the triaxiality of the stress state, which restricts plastic yielding.

6.5.4 Antiplane Strain Case

The plastic zone shape for anti-plane strain, or Mode III case using the elastic K-field is introduced. The result obtained can be compared with that of the complete small-scale yielding solution in Section 6.7. The stresses in the Mode III elastic K-fields are

$$\sigma_{yz} = \frac{K_{III}}{\sqrt{2\pi r}}\cos\frac{1}{2}\theta$$

$$\sigma_{xz} = -\frac{K_{III}}{\sqrt{2\pi r}}\sin\frac{1}{2}\theta$$

Substitution of these stresses into the von Mises yield criterion Eq. (6.5) yields

$$\frac{K_{III}^2}{2\pi r} = \frac{1}{3}\sigma_Y^2$$

Thus, the plastic zone boundary is described by a circle of radius

$$r_p = \frac{3K_{III}^2}{2\pi\sigma_Y^2} \tag{6.61}$$

with the center located at the crack tip. It is of interest to note that the plastic zone in the small-scale yielding solution in Section 6.7 is also a circle with the same radius r_p in Eq. (6.61), but centered at $x = r_p$ along the crack line. The plastic zone size along the crack line is thus twice that given in Eq. (6.61).

6.6 PLASTIC ZONE SHAPE ACCORDING TO FINITE ELEMENT ANALYSES

The plastic zone shapes and sizes estimated from the elastic singular stress field introduced in the previous section are only the first-order approximations. The estimated plastic zone tends to be smaller than the actual size. For example, the plastic zone size given in Eq. (6.55) for plane stress is only half of Irwin's estimate, which is believed to be closer to the actual size. A more realistic estimate of plastic zone shape and size can be made by elastic-plastic finite element analyses. This section introduces the numerical results in Kim [6-4].

Figure 6.10 shows the near-tip plastic zone shapes and sizes for a centrally cracked plate under plane stress conditions using both the K-field and the finite element analysis. The plate is subjected to a tensile load as shown and the crack length is much smaller than the in-plane specimen size (the width of the plate is 10 times the crack length in the calculation). The von Mises yield criterion is used in the calculations. Poisson's ratio is taken as 0.3. For ease of comparison, all the sizes are normalized with respect to that of the elastic solution for the load level of $K_I/(\sigma_Y\sqrt{\pi a}) = 0.3$. It is seen that the plastic zone size on the crack line by the K-field estimate is about half of the finite element solution when the applied stress level $K_I/(\sigma_Y\sqrt{\pi a}) = 0.4$. The difference of the plastic zone size along the y-direction is moderate at this load level.

Figure 6.11 shows the near-tip plastic zones for the plane strain case. It is seen that the plastic zone size on the crack line by the K-field estimate closely matches that by the finite element analysis although the K-field estimate in the y-direction is significantly smaller than that of the numerical analysis even at a load level of $K_I/(\sigma_Y\sqrt{\pi a}) = 0.2$. The plastic zone size along the crack line is less than 5% of the crack length, indicating small-scale yielding prevails at this load level.

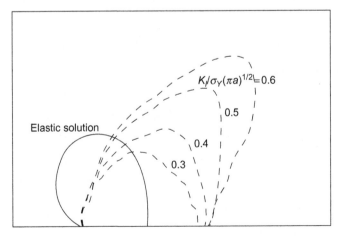

FIGURE 6.10

Plane stress crack tip plastic zone by finite element analyses (adapted from Kim [6-4]).

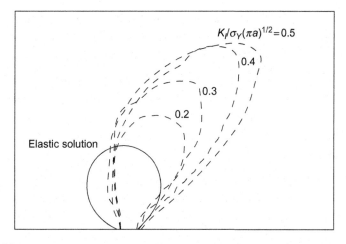

FIGURE 6.11

Plane strain crack tip plastic zone by finite element analysis (adapted from Kim [6-4]).

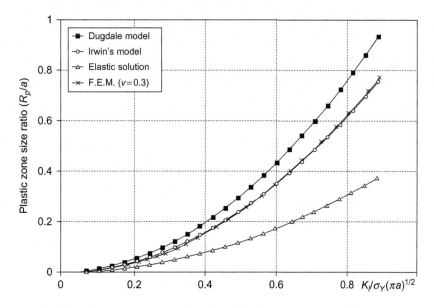

FIGURE 6.12

Plane stress plastic zone size along the crack line for a centrally cracked plate (adapted from Kim [6-4]).

Figure 6.12 further shows the plane stress plastic zone size estimates versus the loading parameter $K_I/(\sigma_Y\sqrt{\pi a})$ along the crack line using the elastic K-field solution, the Dugdale model (based on small-scale yielding), Irwin's model, and the finite element analysis. The plastic zone size is normalized by the half crack length a. It is observed that the plastic zone size predicted from Irwin's model agrees very well with the numerical results. While the K-field solution significantly underestimates the plastic zone size, the Dugdale model generally overestimates the plastic zone size. For the plane strain case, Irwin's prediction is not as good as in the plane stress case, while the elastic solution seems to be quite accurate up to $K_I/(\sigma_Y\sqrt{\pi a}) = 0.5$.

6.7 A MODE III SMALL-SCALE YIELDING SOLUTION

The nonlinearity of the elastic-plastic constitutive relationship makes it extremely difficult to obtain closed-form solutions for elastic-plastic crack problems, especially the in-plane Mode I or Mode II problems. For Mode III crack problems, however, closed-form solutions may be obtained under small-scale yielding conditions. The significance of the Mode III small-scale yielding solution lies in the fact that Irwin's assumption on the elastic field around a Mode I crack in an elastic-perfectly plastic material agrees with the Mode III solution.

This offers some support for Irwin's adjusted stress intensity factor approach based on the effective crack length, which has proven convenient and effective in predicting fracture of metals under small-scale yielding conditions. In this section, the small-scale yielding solution is introduced for an elastic-perfectly plastic material studied by Hult and McClintock [6-5].

6.7.1 Basic Equations

We know from Chapter 3 that, under Mode III deformation conditions, there are only five nonzero field quantities independent of the coordinate z, that is,

$$\tau_x = \sigma_{xz}, \quad \tau_y = \sigma_{yz}$$

$$\gamma_x = 2e_{xz}, \quad \gamma_y = 2e_{yz}$$

$$w = u_z$$

Figure 6.13 shows the coordinate systems used in the study.

The anti-plane shear stresses satisfy the equilibrium equation:

$$\frac{\partial \tau_x}{\partial x} + \frac{\partial \tau_y}{\partial y} = 0 \tag{6.62}$$

The anti-plane shear strains and the out-of-plane displacement are related by

$$(\gamma_x, \gamma_y) = \left(\frac{\partial w}{\partial x}, \frac{\partial w}{\partial y}\right) \tag{6.63}$$

In the elastic region, the preceding equations are supplemented by Hooke's law:

$$(\gamma_x, \gamma_y) = \frac{1}{\mu}(\tau_x, \tau_y) \tag{6.64}$$

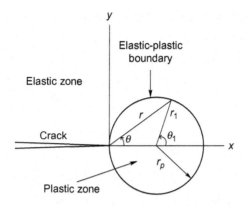

FIGURE 6.13

Plastic zone ahead of a Mode III crack in an elastic-perfectly plastic material in small-scale yielding and coordinate systems.

which is reduced from Eq. (6.6). We thus have five equations in Eqs. (6.62) through (6.64) for the two stresses, two strains, and the displacement.

In the plastic zone, the von Mises yield condition Eq. (6.5) reduces to

$$\tau_x^2 + \tau_y^2 = \tau_Y^2 \tag{6.65}$$

where $\tau_Y = \sigma_Y/\sqrt{3}$ is the yield stress in shear. The constitutive relationship of the deformation plasticity, Eqs. (6.26) and (6.27), now reduces to

$$(\gamma_x, \gamma_y) = \lambda(\tau_x, \tau_y) \tag{6.66}$$

where λ is a proportionality factor. Now we have six equations in Eqs. (6.62), (6.63), (6.65), and (6.66) for the two stresses, two strains, the displacement, and the factor λ.

6.7.2 Elastic-Plastic Solution and the Crack Tip Plastic Zone

In the plastic zone, the yield condition Eq. (6.65) indicates that the stress vector (τ_x, τ_y) has a constant magnitude. It can be easily verified that the stresses

$$\tau_x = -\tau_Y \sin\theta, \quad \tau_y = \tau_Y \cos\theta \tag{6.67}$$

satisfy the equilibrium equation (6.62) and the yield condition Eq. (6.65), where θ is a polar coordinate defined by

$$x = r\cos\theta, \quad y = r\sin\theta \tag{6.68}$$

Substituting the stresses Eq. (6.67) into the constitutive relationship Eq. (6.66) yields the following strains:

$$\gamma_x = -\lambda \tau_Y \sin\theta, \quad \gamma_y = \lambda \tau_Y \cos\theta$$

Using these strain expressions and the strain-displacement relations in Eq. (6.63), we can arrive at

$$dw = \gamma_x dx + \gamma_y dy = 0$$

along $\theta = $ const. ($d\theta = 0$). Hence, the displacement w is a function of θ only, that is,

$$w = F(\theta) \tag{6.69}$$

where $F(\theta)$ is an unknown function to be determined. Use of the displacement above and the strain-displacement relations in the polar coordinates (r,θ) yields the following strains in the polar coordinate system:

$$\gamma_r = \frac{\partial w}{\partial r} = 0, \quad \gamma_\theta = \frac{1}{r}\frac{\partial w}{\partial \theta} = \frac{1}{r}\frac{dF(\theta)}{d\theta} \tag{6.70}$$

It can be seen from Eq. (6.67) that the stresses in the plastic zone do not satisfy the traction-free boundary condition $\tau_y = 0$ on the crack face ($y = 0, x < 0$, or $\theta = \pm\pi$). The plastic zone thus does not contain the crack surfaces. In the small-scale yielding problem, we assume that the yield zone around the crack tip is so small that the elastic K-field is not disturbed at distances far away from the yield zone. Mathematically, this condition can be described by (see Chapter 3 for the Mode III K-field)

$$\tau_y + i\tau_x \to \frac{K_{III}}{\sqrt{2\pi z}}, \quad |z|/r_p \to \infty \tag{6.71}$$

where K_{III} is the Mode III stress intensity factor, $2r_p$ is the yield zone size, and $z = x + iy$ is the complex coordinate. The asymptotic condition at $|z|/r_p \to \infty$ means that we can consider a semi-infinite plane (after symmetry consideration) for the small-scale yielding crack problem, as shown in Figure 6.14.

Outside the plastic zone, we consider the following elastic stresses:

$$\tau_y + i\tau_x = \frac{K_{III}}{\sqrt{2\pi z_1}} = \frac{K_{III}}{\sqrt{2\pi(z - r_p)}} \tag{6.72}$$

where $z_1 = z - r_p$, with r_p being the half plastic zone size along the crack line. The preceding solution clearly satisfies the asymptotic condition Eq. (6.71) and also the crack face traction-free condition because $z_1 \to z$ when $|z|/r_p \to \infty$.

In the following, the stress and displacement continuity conditions are employed to determine the plastic zone shape and size, as well as the function $F(\theta)$ in the

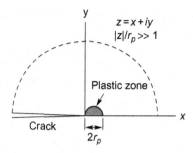

FIGURE 6.14

Small-scale yielding in the physical plane.

displacement Eq. (6.69). First, the elastic stress field Eq. (6.72) and the corresponding displacement are rewritten in polar coordinates (r_1, θ_1) as follows:

$$\left(\tau_x, \tau_y\right) = \frac{K_{III}}{\sqrt{2\pi r_1}} \left(-\sin \frac{\theta_1}{2}, \cos \frac{\theta_1}{2}\right)$$

$$w = \frac{K_{III}}{\mu} \sqrt{\frac{2r_1}{\pi}} \sin \frac{\theta_1}{2}$$

where (r_1, θ_1) is defined by (see Figure 6.13)

$$z_1 = z - r_p = r_1 \left(\cos\theta_1 + i\sin\theta_1\right)$$

Use of the stress and displacement continuity conditions at the elastic-plastic boundary yields the following equations to determine the function $F(\theta)$, r_1, and the relation between θ_1 and θ:

$$\tau_Y \sin\theta = \frac{K_{III}}{\sqrt{2\pi r_1}} \sin \frac{\theta_1}{2}$$

$$\tau_Y \cos\theta = \frac{K_{III}}{\sqrt{2\pi r_1}} \cos \frac{\theta_1}{2} \qquad (6.73)$$

$$F(\theta) = \frac{K_{III}}{\mu} \sqrt{\frac{2r_1}{\pi}} \sin \frac{\theta_1}{2}$$

r_1 and θ_1 can be obtained from the first two equations in Eq. (6.73) as follows:

$$r_1 = \frac{1}{2\pi} \left(\frac{K_{III}}{\tau_Y}\right)^2 \qquad (6.74)$$

$$\theta_1 = 2\theta \qquad (6.75)$$

Hence, the elastic-plastic boundary is a circle with a radius

$$r_p = \frac{1}{2\pi} \left(\frac{K_{III}}{\tau_Y} \right)^2 \tag{6.76}$$

and centered at $(x,y) = (r_p, 0)$, as shown in Figure 6.13.

The function $F(\theta)$, and hence the displacement w in the plastic zone according to Eq. (6.69), can be obtained from the last equation in Eq. (6.73) as follows:

$$w = F(\theta) = \frac{\gamma_Y K_{III}^2}{\pi \tau_Y^2} \sin\theta \tag{6.77}$$

where $\gamma_Y = \tau_Y/\mu$ is the yield strain in shear. Substituting the displacement above into Eq. (6.70), we have the strain components in the plastic zone:

$$\gamma_r = 0, \quad \gamma_\theta = \frac{\gamma_Y K_{III}^2}{\pi \tau_Y^2} \frac{\cos\theta}{r} \tag{6.78}$$

In elastic-plastic fracture mechanics, the CTOD is often used as a fracture parameter. The CTOD, or δ, for the Mode III small-scale yielding problem is obtained using Eq. (6.77) as follows:

$$\delta = w|_{\theta=\pi/2} - w|_{\theta=-\pi/2}$$

$$= \frac{2\gamma_Y}{\pi} \left(\frac{K_{III}}{\tau_Y} \right)^2 \tag{6.79}$$

Some important conclusions can be made from the small-scale yielding solution for the elastic-perfectly plastic material as follows:

1. The plastic zone has a circular shape with a diameter of $(K_{III}/\tau_Y)^2/\pi$ ahead of the crack tip.

2. The stresses are finite and the strains are singular at the crack tip, $r = 0$, with an order of $1/r$ versus $1/\sqrt{r}$ in the linear elastic case.

3. The stresses and displacement in the elastic zone are identical to those in LEFM with the crack tip located at the center of the plastic zone.

The last result is particularly significant because Irwin's adjusted stress intensity factor approach based on the effective crack length agrees with this Mode III solution.

6.8 A MODE III SMALL-SCALE YIELDING SOLUTION—ELASTIC POWER-LAW HARDENING MATERIALS

In the previous section, we introduced the Mode III stress and deformation solutions for a crack in an elastic perfectly plastic material under small-scale yielding (SSY) conditions. Metallic materials usually exhibit strain-hardening behavior. The stress–strain relation for strain-hardening materials may be appropriately described by the power law. This section thus introduces the SSY solution for a Mode III crack in an elastic power-law hardening material obtained by Rice [6-9].

6.8.1 Basic Equations

For elastic power-law hardening materials, the equilibrium equation is still given by Eq. (6.62) and the strain-displacement relations are given by Eq. (6.63). In the hardening material case, we often use the strain compatibility condition by eliminating w in Eq. (6.63) as follows

$$\frac{\partial \gamma_y}{\partial x} - \frac{\partial \gamma_x}{\partial y} = 0 \tag{6.80}$$

In the elastic region, the preceding equations are supplemented by Hooke's law (6.64). We thus have four equations in Eqs. (6.62), (6.80), and (6.64) for the two stresses and two strains. The displacement can be obtained from the strain-displacement relations (6.63) upon solving for the strains.

In the plastic zone characterized by

$$\tau = \sqrt{\tau_x^2 + \tau_y^2} \geq \tau_Y \tag{6.81}$$

the constitutive relationship of the deformation plasticity, Eq. (6.26), reduces to

$$(\gamma_x, \gamma_y) = \frac{\gamma}{\tau}(\tau_x, \tau_y) \tag{6.82}$$

where γ and τ are related by

$$\tau = \tau_Y \left(\frac{\gamma}{\gamma_Y}\right)^N \tag{6.83}$$

in which N is the hardening exponent and γ is

$$\gamma = \sqrt{\gamma_x^2 + \gamma_y^2} \tag{6.84}$$

Now we have four equations in Eqs. (6.62), (6.80), and (6.82) for the two stresses and two strains with τ and γ given in Eqs. (6.81) and (6.84), respectively, and is related by Eq. (6.83).

In the following analytical treatment, we also express the stress and strain vectors in terms of their magnitude and angle. For the stress vector (τ_x, τ_y), we have

$$(\tau_x, \tau_y) = \tau\,(\cos(\phi + \pi/2),\ \sin(\phi + \pi/2)) = \tau(-\sin\phi, \cos\phi) \tag{6.85}$$

where τ is the magnitude of the stress vector and $\phi + \pi/2$ is the angle between the positive x-axis and the vector. It follows from this expression and Eq. (6.82) that

$$(\gamma_x, \gamma_y) = \gamma(-\sin\phi, \cos\phi) \tag{6.86}$$

Here γ is the magnitude of the strain vector (γ_x, γ_y) and $\phi + \pi/2$ is also the angle between the x-axis and the strain vector.

Analytical solutions of the nonlinear plasticity equations described in the preceding are usually not available. For the Mode III SSY crack problem, however, we are able to convert the basic nonlinear equations to linear ones that may admit closed-form solutions. To this end, we do not directly solve for the stresses and strains as functions of coordinates x and y. Instead, we try to solve for x and y as functions of the strains, that is,

$$
\begin{aligned}
x &= x(\gamma_x, \gamma_y) \\
y &= y(\gamma_x, \gamma_y)
\end{aligned}
\tag{6.87}
$$

which, in general, give the strains in implicit forms. Now we need to write all the basic equations in the strain plane instead of the physical plane. Taking the derivatives of both sides in Eq. (6.87) with respective to x, we have

$$
1 = \frac{\partial x}{\partial \gamma_x} \frac{\partial \gamma_x}{\partial x} + \frac{\partial x}{\partial \gamma_y} \frac{\partial \gamma_y}{\partial x}
$$

$$
0 = \frac{\partial y}{\partial \gamma_x} \frac{\partial \gamma_x}{\partial x} + \frac{\partial y}{\partial \gamma_y} \frac{\partial \gamma_y}{\partial x}
$$

Solving these equations for $\partial \gamma_y / \partial x$, we have

$$
\frac{\partial \gamma_y}{\partial x} = -\frac{1}{\Delta} \frac{\partial y}{\partial \gamma_x}
\tag{6.88}
$$

where Δ is the determinant of the Jacobian matrix of the transformation

$$
\Delta = \frac{\partial x}{\partial \gamma_x} \frac{\partial y}{\partial \gamma_y} - \frac{\partial y}{\partial \gamma_x} \frac{\partial x}{\partial \gamma_y}
$$

Δ is not zero if we have one-to-one transformation between the physical plane and the strain plane. Similarly, taking the derivatives of Eq. (6.87) with respect to y leads to

$$
0 = \frac{\partial x}{\partial \gamma_x} \frac{\partial \gamma_x}{\partial y} + \frac{\partial x}{\partial \gamma_y} \frac{\partial \gamma_y}{\partial y}
$$

$$
1 = \frac{\partial y}{\partial \gamma_x} \frac{\partial \gamma_x}{\partial y} + \frac{\partial y}{\partial \gamma_y} \frac{\partial \gamma_y}{\partial y}
$$

Solving the equations for $\partial \gamma_x / \partial y$ yields

$$
\frac{\partial \gamma_x}{\partial y} = -\frac{1}{\Delta} \frac{\partial x}{\partial \gamma_y}
\tag{6.89}
$$

Substituting Eqs. (6.88) and (6.89) into Eq. (6.80), we obtain the compatibility equation in the strain plane

$$\frac{\partial x}{\partial \gamma_y} - \frac{\partial y}{\partial \gamma_x} = 0 \tag{6.90}$$

Similarly, the equilibrium equation, that is, (6.62), may be written in the stress plane using the same procedure described here

$$\frac{\partial x}{\partial \tau_x} + \frac{\partial y}{\partial \tau_y} = 0 \tag{6.91}$$

The compatibility equation (6.90) is automatically satisfied with the following defined strain function $\psi = \psi(\gamma_x, \gamma_y)$

$$x = \frac{\partial \psi}{\partial \gamma_x}, \quad y = \frac{\partial \psi}{\partial \gamma_y} \tag{6.92}$$

These relations may also be written in terms of γ and ϕ using Eq. (6.86)

$$
\begin{aligned}
x &= -\sin\phi \frac{\partial \psi}{\partial \gamma} - \frac{\cos\phi}{\gamma} \frac{\partial \psi}{\partial \phi} \\
y &= \cos\phi \frac{\partial \psi}{\partial \gamma} - \frac{\sin\phi}{\gamma} \frac{\partial \psi}{\partial \phi}
\end{aligned} \tag{6.93}
$$

or

$$\bar{z} = x - iy = re^{-i\theta} = -e^{-i\phi}\left(\frac{1}{\gamma}\frac{\partial \psi}{\partial \phi} + i\frac{\partial \psi}{\partial \gamma}\right) \tag{6.94}$$

Substituting Eq. (6.93) into the equilibrium equation (6.91) and considering the following relation

$$
\begin{aligned}
\frac{\partial ()}{\partial \tau_x} &= -\sin\phi \frac{\partial ()}{\partial \tau} - \frac{\cos\phi}{\tau} \frac{\partial ()}{\partial \phi} \\
\frac{\partial ()}{\partial \tau_y} &= \cos\phi \frac{\partial ()}{\partial \tau} - \frac{\sin\phi}{\tau} \frac{\partial ()}{\partial \phi}
\end{aligned} \tag{6.95}
$$

we have

$$\frac{\tau}{\gamma \tau'(\gamma)} \frac{\partial^2 \psi}{\partial \gamma^2} + \frac{1}{\gamma} \frac{\partial \psi}{\partial \gamma} + \frac{1}{\gamma^2} \frac{\partial^2 \psi}{\partial \phi^2} = 0 \tag{6.96}$$

In the plastic zone ($\tau \geq \tau_Y$, or $\gamma \geq \gamma_Y$),

$$\tau'(\gamma) = N \frac{\tau_Y}{\gamma_Y} \left(\frac{\gamma}{\gamma_Y}\right)^{N-1}$$

Substituting the preceding equation into Eq. (6.96), we obtain the governing equation for the strain function ψ

$$\frac{1}{N} \frac{\partial^2 \psi}{\partial \gamma^2} + \frac{1}{\gamma} \frac{\partial \psi}{\partial \gamma} + \frac{1}{\gamma^2} \frac{\partial^2 \psi}{\partial \phi^2} = 0, \quad \gamma \geq \gamma_Y \tag{6.97}$$

This is a linear partial differential equation for the strain function ψ. In the elastic zone ($\tau < \tau_Y$, or $\gamma < \gamma_Y$),

$$\frac{\partial^2 \psi}{\partial \gamma^2} + \frac{1}{\gamma} \frac{\partial \psi}{\partial \gamma} + \frac{1}{\gamma^2} \frac{\partial^2 \psi}{\partial \phi^2} = 0, \quad \gamma < \gamma_Y \tag{6.98}$$

which is actually a Laplace equation in the strain plane. After solving the previous equations for ψ under appropriate boundary conditions, the relations between γ, ϕ and x, y may be established.

6.8.2 Boundary Conditions of SSY

The SSY condition is still given in Eq. (6.71), or equivalently

$$\gamma_y + i\gamma_x = \frac{1}{\mu}(\tau_y + i\tau_x) \rightarrow \frac{K_{III}}{\mu\sqrt{2\pi z}}, \quad |z|/R_p \rightarrow \infty \tag{6.99}$$

where R_p is a measure of the plastic zone size. The asymptotic condition at $|z|/R_p \rightarrow \infty$ means that we can consider a semi-infinite plane (after symmetry consideration) for the SSY crack problem, as shown earlier in Figure 6.14.

The crack face traction-free condition now becomes

$$\tau_y = 0, \quad y = 0, x < 0 \tag{6.100}$$

and the symmetry condition along the crack line is

$$w = 0, \quad y = 0, x > 0 \tag{6.101}$$

Using Hooke's law (6.64), the elastic-plastic stress–strain relationship (6.82), and the strain-displacement relation (6.63), the conditions in Eqs. (6.100) and (6.101) can be expressed in the following equivalent forms

$$\gamma_y = 0, \quad y = 0, x < 0 \tag{6.102}$$

$$\gamma_x = 0, \quad y = 0, x > 0 \tag{6.103}$$

Because we are dealing with a crack problem, a strain singularity at the tip ($z = 0$) is anticipated, that is,

$$\gamma \rightarrow \infty, \quad z \rightarrow 0 \tag{6.104}$$

Finally, we require that the stresses and strains be continuous across the elastic-plastic boundary, in other words,

$$[[\gamma_x]] = 0, \quad [[\gamma_y]] = 0 \tag{6.105}$$

where $[[\gamma_x]]$ and $[[\gamma_y]]$ represent the discontinuities of γ_x and γ_y across the elastic-plastic boundary, respectively.

Using Eq. (6.86), the boundary conditions in Eqs. (6.103) and (6.102) can be rewritten in terms of ϕ as follows

$$\phi = 0, \qquad y = 0, \, x > 0 \qquad\qquad (6.106)$$

$$\phi = \frac{\pi}{2}, \qquad y = 0, \, x < 0 \qquad\qquad (6.107)$$

respectively. Hence, the physical half plane, $y > 0$, is mapped onto the quarter plane

$$0 < \phi < \pi/2$$

in the strain plane, as shown in Figure 6.15. By noting Eq. (6.93), the symmetry condition in Eq. (6.106) in the strain plane becomes

$$\frac{\partial \psi}{\partial \gamma} = 0, \qquad \phi = 0 \qquad\qquad (6.108)$$

and the traction-free condition (6.107) becomes

$$\frac{\partial \psi}{\partial \phi} = 0, \qquad \phi = \frac{\pi}{2} \qquad\qquad (6.109)$$

The SSY condition in the strain plane follows from Eq. (6.99)

$$z \to \frac{K_{III}^2}{2\pi \mu^2} \frac{1}{(\gamma_y + i\gamma_x)^2}, \qquad \gamma \to 0$$

or by noting Eq. (6.94) and $\gamma_y + i\gamma_x = \gamma e^{-i\phi}$

$$-e^{-i\phi} \left(\frac{1}{\gamma} \frac{\partial \psi}{\partial \phi} + i \frac{\partial \psi}{\partial \gamma} \right) \to \frac{K_{III}^2}{2\pi \mu^2 \gamma^2} e^{-2i\phi}, \qquad \gamma \to 0$$

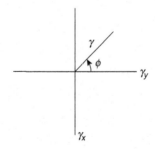

FIGURE 6.15

Strain vector in the strain plane.

The preceding condition can be further simplified as follows

$$\psi \to -\frac{K_{III}^2}{2\pi\mu^2}\frac{\sin\phi}{\gamma}, \quad \gamma \to 0 \tag{6.110}$$

The asymptotic condition at the crack tip (6.104) and the transformation (6.94) imply that the strain function ψ has the asymptotic properties

$$\frac{\partial\psi}{\partial\gamma} \to 0, \quad \frac{1}{\gamma}\frac{\partial\psi}{\partial\phi} \to 0, \quad \gamma \to \infty \tag{6.111}$$

The stress or strain continuity conditions (6.105), together with Eqs. (6.86) and (6.94), mean that both ψ and $\partial\psi/\partial\gamma$ remain continuous across the elastic-plastic boundary (described by $\gamma = \gamma_Y$ in the strain plane)

$$[[\psi]] = 0, \quad \left[\left[\frac{\partial\psi}{\partial\gamma}\right]\right] = 0, \quad \gamma = \gamma_Y \tag{6.112}$$

6.8.3 Elastic-Plastic Solution

The elastic-plastic Mode III crack problem now reduces to finding the solution of governing equations, (6.98) and (6.97), in the elastic and plastic zones, respectively, under the conditions in Eqs. (6.108) through (6.112).

In the elastic region, the SSY condition (6.110) and the boundary conditions (6.108) and (6.109) suggest that the strain function ψ may be expressed in the following separable form

$$\psi(\gamma,\phi) = f(\gamma)\sin\phi \tag{6.113}$$

where $f(\gamma)$ is a function of the magnitude of the strain vector. Substituting this equation into Eq. (6.98) yields the following equation for $f(\gamma)$

$$\frac{d^2f}{d\gamma^2} + \frac{1}{\gamma}\frac{df}{d\gamma} - \frac{f}{\gamma^2} = 0, \quad \gamma < \gamma_Y$$

The general solution of the equation is

$$f(\gamma) = C_1\gamma^{-1} + C_2\gamma$$

where C_1 and C_2 are unknown constants to be determined. The general solution of ψ in the elastic zone thus can be expressed as

$$\psi(\gamma,\phi) = \left(C_1\gamma^{-1} + C_2\gamma\right)\sin\phi, \quad \gamma < \gamma_Y \tag{6.114}$$

In the plastic zone, the strain function ψ may still be expressed in the separable form (6.113) from the consideration of continuity conditions across the elastic-plastic boundary. Substituting ψ from Eq. (6.113) into the basic equation (6.97) yields the following equation for $f(\gamma)$ in the plastic zone

$$\frac{1}{N}\frac{d^2 f}{d\gamma^2} + \frac{1}{\gamma}\frac{df}{d\gamma} - \frac{f}{\gamma^2} = 0, \quad \gamma \geq \gamma_Y$$

with the following general solution

$$f(\gamma) = C_3 \gamma^{-N} + C_4 \gamma$$

where C_3 and C_4 are constants to be determined. The general solution of ψ in the plastic zone can be expressed as

$$\psi(\gamma,\phi) = \left(C_3 \gamma^{-N} + C_4 \gamma\right)\sin\phi, \quad \gamma \geq \gamma_Y \tag{6.115}$$

Clearly, the general solutions in Eqs. (6.114) and (6.115) satisfy the boundary conditions in Eqs. (6.108) and (6.109). We now determine the unknown constants $C_i(i = 1,2,3,4)$ using conditions Eqs. (6.110) through (6.112). It follows from the SSY condition (6.110) and the solution in the elastic zone (6.114) that

$$C_1 = -\frac{K_{III}^2}{2\pi\mu^2} = -\frac{K_{III}^2 \gamma_Y^2}{2\pi\tau_Y^2}$$

The asymptotic condition in Eq. (6.111) and the solution in the plastic zone (6.115) give

$$C_4 = 0$$

According to the continuity requirements (6.112), the remaining two constants C_2 and C_3 satisfy the following conditions after using the ones obtained in C_1 and C_4

$$-\frac{K_{III}^2 \gamma_Y^2}{2\pi\tau_Y^2}\gamma_Y^{-1} + C_2\gamma_Y = C_3\gamma_Y^{-N}, \qquad \frac{K_{III}^2 \gamma_Y^2}{2\pi\tau_Y^2}\gamma_Y^{-2} + C_2 = -NC_3\gamma_Y^{-N-1}$$

Solving these equations for C_2 and C_3, we obtain

$$C_2 = -\frac{K_{III}^2}{2\pi\tau_Y^2}\frac{1-N}{1+N}, \qquad C_3 = -\frac{K_{III}^2}{\pi\tau_Y^2}\frac{\gamma_Y^{N+1}}{(1+N)}$$

Using these determined constants, we have the solution for the strain function ψ as follows

$$\psi(\gamma,\phi) = -\frac{K_{III}^2}{\pi\tau_Y^2}\frac{\gamma_Y}{(1+N)}\left(\frac{\gamma_Y}{\gamma}\right)^N \sin\phi, \quad \gamma \geq \gamma_Y \tag{6.116}$$

$$\psi(\gamma,\phi) = -\frac{K_{III}^2 \gamma_Y}{2\pi\tau_Y^2}\left(\frac{\gamma_Y}{\gamma} + \frac{1-N}{1+N}\frac{\gamma}{\gamma_Y}\right)\sin\phi, \quad \gamma < \gamma_Y \tag{6.117}$$

With the previously solved strain function ψ, we can obtain the stresses and strains for the Mode III SSY elastic-plastic crack problem. Substituting the solution in the plastic zone (6.116) into Eq. (6.94) yields

$$re^{-i\theta} = e^{-i\phi}\frac{K_{III}^2}{\pi\tau_Y^2}\frac{1}{(1+N)}\left(\frac{\gamma_Y}{\gamma}\right)^{N+1}(\cos\phi - iN\sin\phi)$$

$$= \frac{K_{III}^2}{\pi\tau_Y^2}\frac{1}{(1+N)}\left(\frac{\gamma_Y}{\gamma}\right)^{N+1}\left(\frac{1-N}{2} + \frac{1+N}{2}e^{-2i\phi}\right)$$

or

$$x = r\cos\theta = \frac{K_{III}^2}{2\pi\tau_Y^2}\left(\frac{\gamma_Y}{\gamma}\right)^{N+1}\left[\frac{1-N}{1+N} + \cos 2\phi\right]$$

$$\qquad\qquad\qquad\qquad\qquad\qquad\qquad\qquad\qquad (6.118)$$

$$y = r\sin\theta = \frac{K_{III}^2}{2\pi\tau_Y^2}\left(\frac{\gamma_Y}{\gamma}\right)^{N+1}\sin 2\phi$$

The implicit solutions of γ and ϕ in the plastic zone can be obtained from Eq. (6.118) as follows

$$\sin(2\phi - \theta) = \frac{1-N}{1+N}\sin\theta \qquad\qquad (6.119)$$

$$\frac{\gamma}{\gamma_Y} = \left(\frac{K_{III}^2}{2\pi\tau_Y^2}\frac{1}{r}\right)^{\frac{1}{1+N}}\left(\frac{\sin 2\phi}{\sin\theta}\right)^{\frac{1}{1+N}} \qquad\qquad (6.120)$$

Substituting Eq. (6.120) into Eq. (6.86) yields the strains in the plastic zone

$$\gamma_x = -\gamma_Y\left(\frac{K_{III}^2}{2\pi\tau_Y^2}\frac{1}{r}\right)^{\frac{1}{1+N}}\left(\frac{\sin 2\phi}{\sin\theta}\right)^{\frac{1}{1+N}}\sin\phi$$

$$\qquad\qquad\qquad\qquad\qquad\qquad\qquad\qquad\qquad (6.121)$$

$$\gamma_y = \gamma_Y\left(\frac{K_{III}^2}{2\pi\tau_Y^2}\frac{1}{r}\right)^{\frac{1}{1+N}}\left(\frac{\sin 2\phi}{\sin\theta}\right)^{\frac{1}{1+N}}\cos\phi$$

The stresses can be obtained by subsituting Eq. (6.120) into the power law (6.83) and then Eq. (6.85) as follows

$$\tau_x = -\tau_Y\left(\frac{K_{III}^2}{2\pi\tau_Y^2}\frac{1}{r}\right)^{\frac{N}{1+N}}\left(\frac{\sin 2\phi}{\sin\theta}\right)^{\frac{N}{1+N}}\sin\phi$$

$$\qquad\qquad\qquad\qquad\qquad\qquad\qquad\qquad\qquad (6.122)$$

$$\tau_y = \tau_Y\left(\frac{K_{III}^2}{2\pi\tau_Y^2}\frac{1}{r}\right)^{\frac{N}{1+N}}\left(\frac{\sin 2\phi}{\sin\theta}\right)^{\frac{N}{1+N}}\cos\phi$$

In the elastic zone, γ and ϕ may be obtained by substituting Eq. (6.117) into Eq. (6.94) as follows

$$\bar{z} = re^{-i\theta} = \frac{K_{III}^2}{2\pi \tau_Y^2}\left[\left(\frac{\gamma_Y}{\gamma}\right)^2 e^{-2i\phi} + \frac{1-N}{1+N}\right] \tag{6.123}$$

This equation becomes the following after using Eq. (6.86)

$$z - \frac{1-N}{1+N}\frac{K_{III}^2}{2\pi \tau_Y^2} = \frac{K_{III}^2}{2\pi \tau_Y^2}\left(\frac{\gamma_Y}{\gamma}\right)^2 e^{2i\phi} = \frac{K_{III}^2}{2\pi \tau_Y^2}\left(\frac{\gamma_Y}{\gamma_y + i\gamma_x}\right)^2 \tag{6.124}$$

The strains and hence stresses in the elastic zone can thus be obtained as follows

$$\gamma_y + i\gamma_x = \frac{K_{III}}{\mu}\frac{1}{\sqrt{2\pi\left(z - \frac{1-N}{1+N}\frac{K_{III}^2}{2\pi\tau_Y^2}\right)}} \tag{6.125}$$

$$\tau_y + i\tau_x = \frac{K_{III}}{\sqrt{2\pi\left(z - \frac{1-N}{1+N}\frac{K_{III}^2}{2\pi\tau_Y^2}\right)}} \tag{6.126}$$

Finally, the elastic-plastic boundary may be determined by equating γ in Eq. (6.124) to γ_Y as follows

$$z - X_0 = R_0 e^{2i\phi} \tag{6.127}$$

where X_0 and R_0 are given by

$$X_0 = \frac{1-N}{1+N}\frac{K_{III}^2}{2\pi\tau_Y^2}, \qquad R_0 = \frac{K_{III}^2}{2\pi\tau_Y^2} \tag{6.128}$$

Equation (6.127) represents a circle of radius R_0 centered at $x = X_0$ on the x-axis as shown in Figure 6.16. The plastic zone size ahead of the crack is

$$R_p = X_0 + R_0 = \frac{1}{(1+N)\pi}\left(\frac{K_{III}}{\tau_Y}\right)^2 \tag{6.129}$$

In summary, the Mode III small-scale yielding solutions for the elastic power-law material indicate that

1. The plastic zone has a circular shape with a radius of $(K_{III}/\tau_Y)^2/(2\pi)$ centered on the x-axis at $(1-N)(K_{III}/\tau_Y)^2/(2\pi(1+N))$. The crack tip is in the plastic zone.

2. The stresses and strains are singular at the crack tip, $r = 0$, according to $r^{-N/(1+N)}$ and $r^{-1/(1+N)}$, respectively.

3. The solution reduces to that for the elastic perfectly plastic material in Section 6.7 when the hardening exponent $N \to 0$.

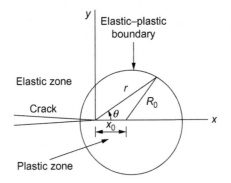

FIGURE 6.16

Plastic zone near the tip of a Mode III crack in an elastic power-law hardening material under small scale yielding.

6.9 HRR FIELD

In elastic-plastic fracture mechanics, a parallel issue to the elastic crack tip K-field is the stress and deformation fields near a crack tip in an elastic-plastic material. Rice and Rosengren [6-6] and Hutchinson [6-7] studied the Mode I problem for a power-law hardening material in the framework of deformation plasticity and obtained the near-tip stress and deformation fields, which are commonly called the HRR field.

Consider a crack in an elastic power-law hardening material as shown in Figure 6.17. The material is assumed to follow the Ramberg-Osgood model in uniaxial loading,

$$\frac{e}{e_Y} = \frac{\sigma}{\sigma_Y} + \alpha \left(\frac{\sigma}{\sigma_Y} \right)^n \tag{6.129}$$

where α is a dimensionless material parameter and n is the hardening exponent. The stress–strain relationship of the deformation plasticity Eqs. (6.26) and (6.27) may be rewritten as

$$e_{ij} = \frac{1+\nu}{E}\sigma_{ij} - \frac{\nu}{E}\sigma_{kk}\delta_{ij} + \frac{3}{2}\frac{\overline{e^p}}{\overline{\sigma}}\sigma'_{ij} \tag{6.130}$$

For the material following the Ramberg-Osgood model Eq. (6.129), Eq. (6.130) reduces to

$$e_{ij} = \frac{1+\nu}{E}\sigma_{ij} - \frac{\nu}{E}\sigma_{kk}\delta_{ij} + \frac{3}{2}\alpha e_Y \left(\frac{\overline{\sigma}}{\sigma_Y} \right)^{n-1} \frac{\sigma'_{ij}}{\sigma_Y} \tag{6.131}$$

Assume stresses have the following asymptotic form near the crack tip:

$$\sigma_{ij} = \sigma_Y A r^\lambda \tilde{\sigma}_{ij}(\theta), \quad r \to 0 \tag{6.132}$$

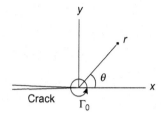

FIGURE 6.17

Coordinate systems at a crack tip.

where A is a constant, $\lambda < 0$ is a singularity exponent, $\tilde{\sigma}_{ij}(\theta)$ are the angular distributions, and (r,θ) are the polar coordinates centered at the crack tip, see Figure 6.17. Substituting the previous asymptotic expression into the constitutive law Eq. (6.131), we see that the plastic strain has a stronger singularity than the elastic strain, which thus can be neglected in the near-tip singular fields. Equation (6.131) is then reduced to the following simpler form in the near-tip region:

$$e_{ij} = \frac{3}{2}\alpha e_Y \left(\frac{\bar{\sigma}}{\sigma_Y} \right)^{n-1} \frac{\sigma'_{ij}}{\sigma_Y} \tag{6.133}$$

and the strains have the asymptotic form near the crack tip:

$$e_{ij} = \alpha e_Y A^n r^{\lambda n} \tilde{e}_{ij}(\theta), \qquad r \to 0 \tag{6.134}$$

To determine the singularity exponent λ, first evaluate the strain energy density W near the crack tip. It follows from Eqs. (6.132) through (6.134) that

$$W = \int_0^{e_{ij}} \sigma_{ij} de_{ij} = \alpha \sigma_Y e_Y \frac{n}{n+1} \left(\frac{\bar{\sigma}}{\sigma_Y} \right)^{n+1} \to r^{\lambda(n+1)} \tilde{W}(\theta), \quad r \to 0 \tag{6.135}$$

Recall that the J-integral is the energy release rate for nonlinear elastic materials and the stress–strain relation Eq. (6.133) is essentially a kind of nonlinear elastic constitutive law. Evaluating J along a circle surrounding the crack tip, we have

$$J = r \int_{-\pi}^{\pi} \left[W \cos\theta - \sigma_{ij} n_j \frac{\partial u_i}{\partial x} \right] d\theta \tag{6.136}$$

For J to be nonzero and finite, the integrand in this integral should have a singularity of $1/r$ at the crack tip, that is,

$$W \to \frac{\tilde{W}(\theta)}{r}, \qquad r \to 0 \tag{6.137}$$

The singularity exponent λ is thus determined by observing Eqs. (6.135) and (6.137):

$$\lambda = -\frac{1}{n+1}$$

Substituting this exponent into Eqs. (6.132) and (6.134) yields the following near-tip stress and strain fields:

$$\sigma_{ij} = \sigma_Y A r^{-\frac{1}{n+1}} \tilde{\sigma}_{ij}(\theta), \qquad r \to 0 \tag{6.138}$$

$$e_{ij} = \alpha e_Y A^n r^{-\frac{n}{n+1}} \tilde{e}_{ij}(\theta), \qquad r \to 0 \tag{6.139}$$

The stress and strain angular distribution functions $\tilde{\sigma}_{ij}$ and $\tilde{e}_{ij}$ in Eqs. (6.138) and (6.139) may be obtained by using the equilibrium equations, the strain compatibility condition, the stress–strain relation in Eq. (6.133), and the appropriate boundary conditions; see Rice and Rosengren [6-6] and Hutchinson [6-7]. These angular distributions also depend on the hardening exponent n. The amplitude constant A can be determined by substituting the asymptotic stresses and strains in Eqs. (6.138) and (6.139) into Eq. (6.136) with the result

$$J = \alpha \sigma_Y e_Y A^{n+1} I_n$$

where I_n is a known constant depending on the hardening exponent n. The final forms of the HRR field can be expressed as follows:

$$\sigma_{ij} = \sigma_Y \left(\frac{J}{\alpha \sigma_Y e_Y I_n r} \right)^{\frac{1}{n+1}} \tilde{\sigma}_{ij}(\theta, n), \quad r \to 0 \tag{6.140}$$

$$e_{ij} = \alpha e_Y \left(\frac{J}{\alpha \sigma_Y e_Y I_n r} \right)^{\frac{n}{n+1}} \tilde{e}_{ij}(\theta, n), \quad r \to 0 \tag{6.141}$$

$$u_i = \alpha e_Y \left(\frac{J}{\alpha \sigma_Y e_Y I_n} \right)^{\frac{n}{n+1}} r^{\frac{1}{n+1}} \tilde{u}_i(\theta, n), \quad r \to 0 \tag{6.142}$$

The HRR field shows that (1) the stresses have a singularity of $r^{-1/(n+1)}$ at the crack tip in a power-law hardening material, and (2) the intensity of the singular stress field is characterized by the J-integral.

6.10 ENERGY RELEASE RATE CONCEPT IN ELASTIC-PLASTIC MATERIALS

Energy release rate plays a very important role in Griffith's theory of fracture. It is a common notion that a crack extends because of the surplus energy released to provide the energy needed in creating new surfaces during crack growth. From the energy balance point of view, we have accepted that, during crack extension of Δa, the work done by external forces ΔW_e is equal to the sum of the increase of elastic strain energy ΔU and the energy released ΔW_s. However, within the framework of classical elasticity, there is no mechanism that is capable of storing surface energy. Therefore, released energy should be absent from the body after crack extension, and, as a result, the energy balance requirement during crack extension cannot be satisfied.

Extending the Griffith energy balance concept for fracture in elastic-plastic solids, we have

$$\Delta W_e = \Delta U + \Delta W_s + \Delta W_p \tag{6.143}$$

in which ΔW_p is the increment of plastic dissipation work. Using the finite element analysis, Sun and Wang [6-8] showed that the energy balance (Eq. 6.143) is satisfied with $\Delta W_s = 0$. They consider a rectangular panel containing a center crack subjected to uniform tension. The applied stress is kept constant during crack extension. A state of plane stress is assumed, and the four-node plane stress element is used to model the panel. For low loading levels, finer meshes are used to capture the small plastic zone.

The material is elastic-perfectly plastic following the von Mises yield criterion and the associated flow rule. Unloading is elastic following the initial modulus of elasticity. The material constants are given by Young's modulus $E = 207$ GPa, Poisson's ratio $\nu = 0.32$, and the yield stress $\sigma_Y = 310$ MPa.

As shown in Figure 6.18, the crack tip is to grow from position A to position B by an amount Δa. To simulate the continuous crack growth process, essentially a large number of extension steps must be performed from A to B. In other words, each numerical nodal release step δa must be sufficiently small to capture the work done during the continuous advance of the plastic zone ahead of the crack tip. The plastic dissipation work is underestimated if the extension step is not sufficiently small.

Figure 6.19 shows the relationships between ratios of energy rates versus $\delta a / R_p$. The energy rates are defined in a manner similar to the enegy release rate, by

$$G_s = \Delta W_s / \Delta a$$

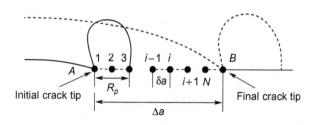

FIGURE 6.18

Crack extension in an elastic-plastic material (Δa is the total growth length and δa is the growth length per extension step).

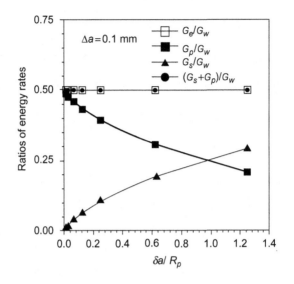

FIGURE 6.19

Ratios of energy rates versus $\delta a/R_p$ (adapted from Sun and Wang [6-8]).

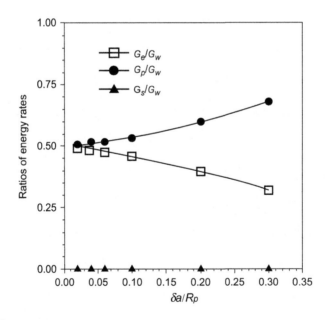

FIGURE 6.20

Effects of plastic zone size on the ratios of energy rates (adapted from Sun and Wang [6-8]).

and

$$G_w = \Delta W_e / \Delta a$$

$$G_p = \Delta W_p / \Delta a$$

$$G_e = \Delta U / \Delta a$$

It is found that G_s approaches zero and G_p approaches the value of G_e as $\delta a / R_p$ decreases. Of course, theoretically the converged solution is obtained with $\delta a \to 0$. These results indicate that, according to the continuum elastoplasticity model, there is no surplus energy released during crack extension and that the only mechanism for energy dissipation during crack extension is plastic work.

It is of interest to investigate the result taking the elastic solution as a limiting case of the elastic-plastic solution. This may be achieved by greatly raising the value of the yield stress, or making the crack tip plastic zone size very small. The finite element method results are shown in Figure 6.20. First, we note that the separation work (energy release) vanishes independent of the plastic zone size. As the plastic zone size increases, the plastic dissipation work rate increases while the elastic strain energy variation rate decreases.

It is interesting to note that, as R_p approaches zero, both G_e and G_p approach 1/2. This means that the elastic energy release rate G_s, which is equal to G in LEFM, is equal to the plastic dissipation work rate when an elastic solid is considered as a limiting case of an elastic-perfectly plastic solid by taking $R_p \to 0$. Thus, the difference between G_p and G may be used as a quantitative measure of the validity of the small-scale yielding concept.

References

[6-1] G.R. Irwin, Plastic zone near a crack and fracture toughness, in: Proceedings of the 7th Sagamore Ordnance Materials Conference, Syracuse University, Syracuse, New York, 1960, pp. IV-63–IV-78.

[6-2] D.S. Dugdale, Yielding of steel sheets containing slits, J. Mech. Phys. Sol. 8 (1960) 100–104.

[6-3] F.M. Burdekin, D.E.W. Stone, The crack opening displacement approach to fracture mechanics in yielding materials, J. Strain Anal. 1 (1966) 145–153.

[6-4] H-O. Kim, Elastic-plastic fracture analysis for small scale yielding, PhD thesis, School of Aeronautics and Astronautics, Purdue University, West Lafayette, IN, 1996.

[6-5] J.A.H. Hult, F.A. McClintock, Elastic-plastic stress and strain distributions around sharp notches under repeated shear, in: Proceedings of the 9th International Congress for Applied Mechanics, Vol. 8, 1957, pp. 51–58.

[6-6] J.R. Rice, G.F. Rosengren, Plane strain deformation near a crack tip in a power-law hardening material, J. Mech. Phys. Sol. 16 (1968) 1–12.

[6-7] J.W. Hutchinson, Singular behavior at the end of a tensile crack in a hardening material, J. Mech. Phys. Sol. 16 (1968) 13–31.

[6-8] C.T. Sun and C.Y. Wang, A new look at energy release rate in fracture mechanics, Inter. J. Fract. 113 (2002) 295–307.

[6-9] J.R. Rice, Stresses due to a sharp notch in a work-hardening elastic-plastic material loaded by longitudinal shear, J. Appl. Mech. 34 (1967) 287–298.

PROBLEMS

6.1 A split beam of 2 mm wide is loaded as shown in Figure 6.21. Assume that the yield stress of the material is 500 MPa. Find the load P that produces a plastic zone size of 2 mm ahead of the crack tip, using the methods that are available to you, given Young's modulus $E = 70$ GPa.

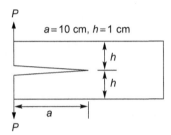

FIGURE 6.21

A split beam of elastic-plastic material.

6.2 Consider an infinite plate with a center crack subjected to a pair of concentrated forces as shown in Figure 3.8 in Chapter 3. Estimate the plastic zone sizes using Irwin's plastic zone adjustment method and the Dugdale model based on

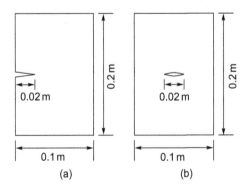

FIGURE 6.22

(a) An edge-cracked plate, and (b) a center-cracked plate.

the small-scale yielding assumption. Also derive the plastic zone size using the Dugdale model without imposing the small-scale yielding condition.

6.3 Consider an edge-cracked plate and a center-cracked panel as shown in Figure 6.22(a) and (b), respectively. Assume that both plates are subjected to a uniform tensile stress applied in the vertical direction. Use finite element analysis to determine the plastic zone size for various levels of the applied load. Assume elastic-perfectly plastic stress–strain behavior. (a) Plot the plastic zone according to plane stress and plane strain assumptions and compare with the results obtained using the elastic solutions. (b) Compare the plastic zone size (along the crack plane) at different load levels with those estimated according to Irwin's approach.

Elastic-Plastic Fracture Criteria

We have learned from Sections 3.9 and 4.1 that (quasistatic) crack growth in perfectly elastic materials can be predicted by the stress intensity factor criterion or equivalently the energy release rate criterion. Under a narrowly defined small-scale yielding (SSY) condition, that is, the plastic deformation is confined in the K-dominance zone of the linear elastic fracture mechanics (LEFM), the stress intensity factor or energy release rate can be used to predict fracture in elastic-plastic materials. The high fracture toughness for metals (Table 3.1 in Section 3.9) indeed results from the energy dissipation in the small plastic zone. When the "strict" SSY condition is not satisfied but the plastic deformation is still confined in the crack tip region, the LEFM approach may still be used by adopting Irwin's adjusted stress intensity factor concept.

Prediction of quasistatic crack growth (including crack initiation and extension) has not been as successful as that in LEFM when large-scale yielding (LSY) conditions prevail, that is, the size of the plastic zone at crack initiation and during crack growth is comparable to, or larger than, the crack length or other in-plane dimensions. In the case of LSY, parameters that describe the overall plastic deformation around the crack tip are needed for crack growth prediction. These parameters include the crack opening displacement (COD), crack tip opening angle (CTOA), and others. This chapter introduces various fracture criteria for predicting quasistatic crack growth in materials undergoing plastic deformations.

7.1 IRWIN'S ADJUSTED STRESS INTENSITY FACTOR APPROACH

Irwin's model introduced in Section 6.3 suggests that the adjusted stress intensity factor with the plastic zone size may be used to predict fracture in elastic-plastic materials under moderate yielding conditions. For this reason, Irwin's approach is also introduced in the LEFM part of most fracture mechanics books. To introduce Irwin's adjusted stress intensity factor approach, we first look at, as shown in Figure 7.1 [7-1], the stress distributions σ_{yy} along the crack line in a center-cracked plate (plane stress) as predicted from Irwin's model and the finite element calculations for an elastic-perfectly plastic material.

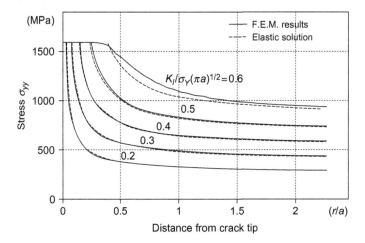

FIGURE 7.1

Stress distribution in front of the crack as predicted from Irwin's model and the finite element method (adapted from Kim [7-1]).

It can be seen that the stress of Irwin's model closely matches the finite element results for load levels up to $K_I/(\sigma_Y\sqrt{\pi a}) = 0.5$, implying that Irwin's model may be used beyond the SSY limitations. Here, K_I is the applied stress intensity factor, σ_Y is the yield stress, and a is half the crack length. Because the stress field around the crack tip in Irwin's model is characterized by the adjusted stress intensity factor, $K_I(a_{eff})$, using the effective crack length a_{eff}, the crack growth condition may be formulated as follows:

$$K_I(a_{eff}) = K_c \tag{7.1}$$

where a_{eff} is the effective crack length given by Eq. (6.38) in Chapter 6 and K_c is a material constant. This K_c should not be confused with the fracture toughness discussed in Chapter 3. This K_c also depends on the thickness of the material.

Irwin's adjusted stress intensity approach has often been labeled in the literature as an SSY approach. The SSY has been frequently referred to and used in general fracture mechanics studies. Hence the SSY concept merits further discussion. The SSY usually means that the crack tip plastic zone is so small that the stress intensity factor in LEFM is applicable. Based on this understanding, the K-dominance zone exists near a crack tip and the plastic zone extends to only a fraction of the K-dominance zone size, as schematically shown in Figure 7.2. The SSY Mode III solution in Section 6.7 is studied based on this assumption.

A broadly defined SSY, or an extension of the previous narrowly defined SSY, may originate from Irwin's model. In this case, the stress intensity factor is still applicable, but modified by using Irwin's effective crack length. Now the plastic zone size may not be negligible compared with the crack length, or other in-plane dimensions.

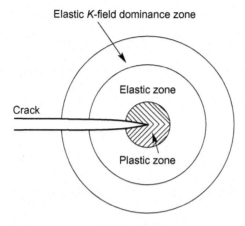

Elastic K-field dominance zone

Elastic zone

Crack

Plastic zone

FIGURE 7.2

Schematic of crack tip SSY.

The stress distributions σ_{yy} shown in Figure 7.1 indicate that Irwin's model may be used for materials undergoing moderate plastic yielding. Irwin's adjusted stress intensity factor is generally larger than the stress intensity in LEFM. For a crack in an infinite plate under uniform remote tension, for example, the adjusted stress intensity factor is given by Eq. (6.39):

$$K_I(a_{eff}) = \sigma\sqrt{\pi a}\Big/\sqrt{1 - \frac{1}{2}\left(\frac{\sigma}{\sigma_Y^*}\right)^2}$$

7.2 *K* RESISTANCE CURVE APPROACH

The K_R curve approach is also commonly characterized as an LEFM fracture criterion as the stress intensity factor K is employed. The K_R curve behavior, however, is a result of increased energy dissipation in the plastic zone that grows with crack extension. If we measure the stress intensity factor K for an elastic-plastic material (especially a thin sheet) with a crack, we will observe that the stress intensity factor increases with crack growth as shown in Figure 7.3, where a_0 is the initial crack size, $a - a_0$ is the crack extension, and K_i is the critical stress intensity factor at crack initiation. The crack growth in the material is thus stable before the final failure occurs.

This is the so-called K resistance curve behavior, or simply K_R curve behavior. Materials, particularly thin-sheet materials, with high fracture toughness and low-yield strength, exhibit K_R curve behavior because a relatively larger plastic zone develops around the crack tip. Now the fracture resistance force is characterized by

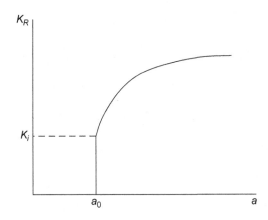

FIGURE 7.3

K resistance curve.

a curve of K_R versus crack extension instead of a single parameter K_c. It is obvious that the resistant curve can be expressed alternately in energy release rate G.

For a material exhibiting K_R curve behavior, failure is determined not only by the comparison between the applied K and K_R, but also by the comparison between the slopes of the applied K and K_R curve. Unstable crack growth takes place when the following two conditions are met:

$$K_I(a) = K_R(\Delta a),$$
$$\frac{dK_I(a)}{da} = \frac{dK_R(\Delta a)}{d(\Delta a)} \tag{7.2}$$

The first equation in Eq. (7.2) is equivalent to Eq. (7.1) but is taken at the current crack length. The second equation in Eq. (7.2) implies that the applied K will continue to exceed K_R for a small crack extension from the current crack length (initial length plus crack extension).

Figure 7.4 shows the K_R curve and a set of fracture driving force curves at different load levels for a given cracked specimen with an initial flaw size a_0. No crack growth occurs at load level P_1 because the applied K is always lower than the K_R for the crack length considered. The crack initiation takes place when the load reaches P_2, as $K(P_2) = K_R(a_0)$ at $a = a_0$. The crack growth, however, is stable as the slope of the driving force curve is smaller than that of the K_R curve at $a = a_0$. The crack extension stops, as $K(P_2)$ is lower than K_R for longer cracks. At the load level P_3, unstable crack growth occurs because both conditions in Eq. (7.2) are met.

It is important to note that condition Eq. (7.2) holds only for the SSY cases, that is, a K-dominance zone exists and is characterized by the stress intensity factor based on the physical crack length. When the SSY condition is not satisfied, Irwin's adjusted

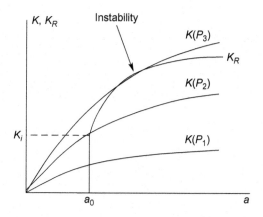

FIGURE 7.4

Fracture instability determined from the driving and resistance curves.

crack length a_{eff} should be used and the criterion Eq. (7.2) now becomes

$$K_I(a_{eff}) = K_R(\Delta a_{eff}),$$
$$\frac{dK_I(a_{eff})}{da_{eff}} = \frac{dK_R(\Delta a_{eff})}{d(\Delta a_{eff})} \qquad (7.3)$$

Bray et al. [7-2] studied the K_R curves for an aluminum alloy (C188-T3) using center-cracked tension (or middle-crack tension, M(T)) specimens as shown in Figure 7.5. The K_R curve may be determined from

$$K_R(\Delta a_{eff}) = \frac{P}{BW}\sqrt{\pi a_{eff}}F\left(\frac{2a_{eff}}{W}\right), \quad F(\alpha) = 1 + 0.128\alpha - 0.288\alpha^2 + 1.525\alpha^3$$

where P is the applied load corresponding to the crack extension, W is the width, B is the thickness, $2a$ is the physical crack length, and $2a_{eff}$ is the effective crack length.

Figure 7.6 shows the K_R versus effective crack growth curves for aluminum specimens with a width of 6.3, 16, and 60 inches, respectively. It can be seen that the K_R curves are basically independent of the specimen size, which indicates that the K_R curve in terms of the effective crack extension is a material property. Of course, the K_R curve is thickness dependent. Figure 7.7 shows the K_R curves based on the physical crack growth for the aluminum specimens. The K_R curve now depends on the specimen size, that is, different specimen sizes result in different K_R curves. Hence, the K_R curve is not a material property if it is based on the physical crack length. The reason is that the plastic zone is not small in this case, which invalidates the use of physical crack length.

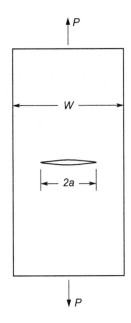

FIGURE 7.5

A center-cracked specimen in tension.

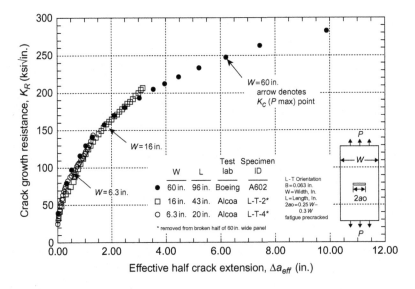

FIGURE 7.6

K_R curves for C188-T3 aluminum with different specimen sizes using the effective crack length (adapted from Bray et al. [7-2]).

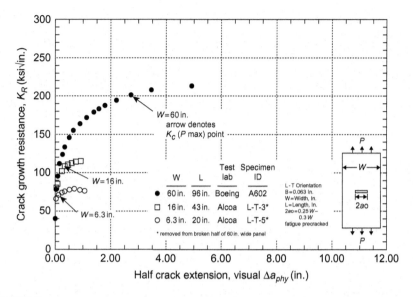

FIGURE 7.7

K_R curves for a center-cracked specimen with different specimen sizes using the physical crack length (adapted from Bray et al. [7-2]).

7.3 *J*-INTEGRAL AS A FRACTURE PARAMETER

The HRR field in Eqs. (6.140) through (6.142) in Section 6.9 of Chapter 6 indicates that the *J*-integral describes the intensity of the near-tip singular stress and deformation fields for a power-law hardening material similar to the way that the stress intensity factor describes the stress and strain intensity in linear elastic materials. A natural extension of the stress intensity factor methodology to the elastic-plastic fracture mechanics is to use *J* as a fracture parameter. Begley and Landes [7-3] proposed the following elastic-plastic crack initiation condition based on this concept:

$$J = J_{Ic} \tag{7.4}$$

where J_{Ic} is the critical value of *J* at crack initiation for plane strain. J_{Ic} should be a material property and determined by experiment. Under SSY conditions, we have the relation between *J* and stress intensity factor (plane strain Mode I):

$$J = \frac{1 - v^2}{E} K_I^2$$

Thus, the *J* criterion is equivalent to the stress intensity factor criterion in SSY.

Under LSY conditions, however, applications of *J* to predicting elastic-plastic fracture have been limited due to several factors. First, the HRR field is based on the deformation theory of plasticity. Deformation plasticity generally requires proportional loading and thus applies to cracked bodies subjected to monotonically increasing loads. Once crack growth occurs, unloading takes place in the wake region along the crack faces, which invalidates the deformation plasticity. As a result, J may be used only for predicting crack initiation but not crack growth in general. Some studies (e.g., Hutchinson and Paris [7-4]), however, suggested that J may be used to predict crack extension if the extension amount is small so that an HRR-dominance zone still exists.

Second, for J to be the sole parameter to describe the stress and deformation intensity near a stationary crack tip, the HRR singular solution must dominate the stress field there. In ductile fracture of metals, large deformations are significant around the crack tip as evidenced by crack tip blunting. The HRR-dominance zone must not fall into the region where large deformations are significant because the small deformation assumption is used in establishing the HRR field. The dominance of the HRR field depends on strain hardening properties as well as configurations of cracked specimens.

The dominance of the HRR field deteriorates significantly for weakly hardening materials. When $n \gg 1$, the stress singularity exponent $-1/(n+1)$ becomes very small, which allows other terms in the asymptotic stress solution to be comparable to the HRR stress field. Finite element analyses of the crack tip deformation (e.g., McMeeking and Parks [7-5] and Shih and German [7-6]) for plane strain showed that the HRR field exists for the single-edge notched bend and double-edge notched tension specimens with hardening exponent $n = 10$ when the following condition is approximately met:

$$\frac{b\sigma_Y}{J} \geq 30 \qquad (7.5)$$

where b is the ligament size and σ_Y is the yield stress. For the single-edge notched tension or center-cracked tension specimens with $n = 10$, however, the HRR field may exist only when

$$\frac{b\sigma_Y}{J} \geq 200 \qquad (7.6)$$

Inequalities Eqs. (7.5) and (7.6) mean that the J-integral may be used to predict crack initiation for LSY under some restrictive conditions.

7.4 CRACK TIP OPENING DISPLACEMENT CRITERION

In LEFM, the near-tip stress and strain fields are always characterized by the inverse square-root singularity and the stress intensity factor is a fracture parameter for both crack initiation and propagation. In elastic-plastic fracture mechanics, however, it is much more difficult to find a similar parameter directly characterizing the intensity of the near-tip singular deformation field for both stationary and growing cracks under LSY conditions. For example, the K_R curve concept based on the effective crack length may not be applied when the plastic zone touches the specimen boundary. Also, the J-integral theoretically holds only for stationary cracks.

On the other hand, parameters describing the near-tip crack profile have been suggested for predicting fracture because the crack profile near the tip reflects the overall severity of the plastic deformation. The crack tip opening displacement (CTOD) and the crack tip opening angle (CTOA) approaches have gained much attention in recent years and proved promising as fracture parameters for both crack initiation and extension prediction. This section describes the CTOD concept and the CTOA approach is discussed in the next section.

According to the CTOD method (Wells [7-7]), the Mode I crack initiation occurs once the the following condition is met:

$$\delta = \delta_c \tag{7.7}$$

where δ is the CTOD, and δ_c is the critical value of the CTOD and is determined by experiments. Once the CTOD is calculated for a cracked structure, Eq. (7.7) may be used to predict the failure load for a given crack size.

Calculation of CTOD according to Irwin's model was discussed in Section 6.3. The result is

$$\delta = \frac{4}{\pi E^*} \frac{K_I^2}{\sigma_Y^*} \tag{7.8}$$

where $E^* = E$ and $\sigma_Y^* = \sigma_Y$ for plane stress, and $E^* = E/(1 - \nu^2)$ and $\sigma_Y^* = \sigma_Y/(1 - 2\nu)$ for plane strain. A relation between δ and K_I was also established using the Dugdale model in SSY in Section 6.4:

$$\delta = \frac{K_I^2}{E^* \sigma_Y^*} \tag{7.9}$$

For SSY, we may have a general relationship between the CTOD and the stress intensity factor as follows:

$$\delta = \beta \frac{K_I^2}{E^* \sigma_Y^*} \tag{7.10}$$

where β is a parameter of order unity equal to 1 according to the Dugdale model and $4/\pi$ according to Irwin's model. The relation between the CTOD and the stress intensity factor indicates that the CTOD criterion is equivalent to the stress intensity criterion in SSY.

Under LSY conditions, the CTOD may be obtained in closed form using the Dugdale model. Otherwise, the finite element method is used to obtain the crack tip opening displacement value. For a crack in an infinite plate subjected to remote tension σ, the CTOD according to the Dugdale model is given by

$$\delta = -\frac{8a\sigma_Y^*}{\pi E^*}\ln\left(\cos\frac{\pi\sigma}{2\sigma_Y^*}\right) \tag{7.11}$$

Using Eqs. (7.7) and (7.11), the crack extension condition becomes

$$-\frac{8a\sigma_Y^*}{\pi E^*}\ln\left(\cos\frac{\pi\sigma}{2\sigma_Y^*}\right) = \delta_c$$

For a given crack length $2a$, the failure applied stress is obtained from the preceding equation:

$$\sigma_f = \frac{2}{\pi}\sigma_Y^*\cos^{-1}\left[\exp\left(-\frac{\pi E^*\delta_c}{8a\sigma_Y^*}\right)\right]$$

For a given applied load, the allowable crack length becomes

$$2a = -\frac{\pi E^*\delta_c}{4\sigma_Y^*}\Big/\ln\left(\cos\frac{\pi\sigma}{2\sigma_Y^*}\right)$$

In the Dugdale or Irwin's model, the CTOD can be determined unambiguously as the crack opening at the physical crack tip. For general ductile fracture problems, a proper definition of CTOD is required. For ductile materials, significant crack tip blunting occurs at crack initiation. Figure 7.8 shows a blunted crack tip region versus the initially sharp crack tip. The CTOD by definition is the crack opening displacement at the orginal sharp crack tip, as shown in Figure 7.8. This definition, however, is not convenient for numerical computations.

Tracy [7-8] proposed a 45° interception definition of CTOD for stationary cracks, as shown in Figure 7.9. The CTOD is now the crack opening at the point at which the crack faces intercept the straight lines of 45° emanating from the blunted crack tip. Using the 45° interception definition and the HRR field, Shih [7-9] derived a relation between the J-integral and CTOD as follows:

$$\delta = d_n\frac{J}{\sigma_Y}$$

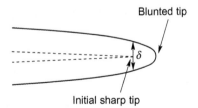

FIGURE 7.8

CTOD definition based on the original crack tip position.

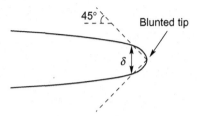

FIGURE 7.9

CTOD definition based on the 45° intercept assumption.

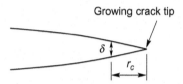

FIGURE 7.10

CTOD definition based on a characteristic distance r_c.

where d_n is a constant depending on the yield strain and the power-hardening exponent n. It should be noted that this relation is valid only when the HRR solution dominates the crack tip stress and deformation fields.

For growing cracks, the crack tip blunting is not significant. Theoretical analyses of near-tip fields around a growing crack in a perfectly plastic material (Rice [7-10]) show that the opening at the growing crack tip vanishes. In this case, the following modified CTOD criterion may be used:

$$\delta = \delta_c, \quad \text{at } r = r_c$$

where r_c is a characteristic distance from the crack tip, as shown in Figure 7.10. In engineering applications, r_c may be calibrated by matching the predicted crack growth response using the preceding criterion with the experimentally measured one. It will be seen in the following section that the CTOD criterion based on the length parameter r_c is actually equivalent to the CTOA criterion.

7.5 CRACK TIP OPENING ANGLE CRITERION

Similar to CTOD, the CTOA is also a parameter describing the overall severity of plastic deformation in the near-tip region. Theoretically, CTOA is defined as the angle between the two crack faces at the crack tip:

$$\text{CTOA} = \psi = 2\tan^{-1}\left(\frac{\delta_0}{2r_0}\right) \tag{7.12}$$

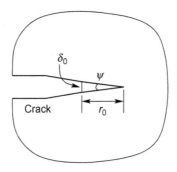

FIGURE 7.11

CTOA definition.

FIGURE 7.12

Crack profile in an aluminum alloy.

where δ_0 is the crack opening displacement at distance r_0 behind the crack tip, as shown in Figure 7.11. If the crack faces follow perfect straight lines in the near-tip region, the CTOA will not depend on the selection of r_0. In real materials, however, the crack faces follow meandering paths. In this case, the CTOA is defined as the average value of the crack opening angle over a small distance behind the crack tip. Figures 7.12 and 7.13 show the crack profiles in the near-tip region for an aluminum alloy and a duplex steel, respectively. It can be seen from the figure that CTOA can be well-defined for the crack growth in the steel, but an averaged CTOA must be used for the aluminum alloy.

According to the CTOA criterion, crack extension occurs when

$$\text{CTOA} = \psi_c \tag{7.13}$$

where ψ_c is the critical CTOA and is determined by experiment.

FIGURE 7.13

Profile of a crack after growth of 5 mm in a duplex steel.

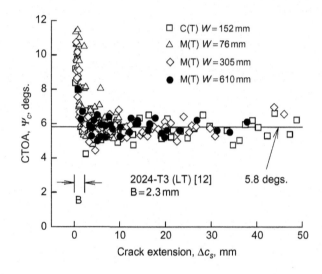

FIGURE 7.14

Measured CTOA data for crack growth in thin sheet, aluminum 2024-T3 specimens (adapted from Newman et al. [7-12]).

The CTOA criteron has gained much attention in recent years in the study of fracture processes in ductile materials (Newman and Zerbst [7-11]). To simulate ductile fracture using the CTOA criterion, the critical CTOA, ψ_c, in Eq. (7.13) needs to be experimentally determined. Figure 7.14 shows the measured critical CTOA values versus crack extension on the surface, Δc_s, for a thin-sheet aluminum alloy (Newman et al. [7-12]). Results for both compact tension (C(T)) and middle-crack tension (M(T)) specimens are included. The specimens have different widths (W) but the same thickness of 2.3 mm.

It can be observed from the figure that (1) the critical CTOA at crack initiation is much higher than that during crack extension; (2) the critical CTOA decreases rapidly with increasing crack extension at the initial growth stage and approximately reaches a constant value of about 5.8 degrees after a small amount of crack growth; and (3) the critical CTOAs for different specimens basically fall in a narrow band, which indicates the existence of a CTOA versus crack extension curve (CTOA resistance curve) that is independent of specimen size and geometry. Hence, the CTOA curve may be regarded as a material property.

Elastic-plastic fracture criteria typically are thickness-dependent, that is, the critical value of the fracture parameter varies with specimen thickness. This has been known for the criterion based on Irwin's model of adjusted stress intensity factor. Mahmound and Lease [7-13] measured the critical CTOA values on the surface of C(T), 2024-T351 aluminum alloy specimens with thicknesses from 2.3 to 25.4 mm. All the specimens have the same width of 203 mm. Their results confirmed the basic characteristic of CTOA curve behavior for all specimens of different thicknesses, that is, for a given specimen, the CTOA has a high value at crack initiation, decreases rapidly and reaches a constant value after a small amount of crack extension. The CTOA curve, however, depends on the specimen thickness.

Figure 7.15 shows the variation of critical CTOA values in the constant region with specimen thickness (Mahmound and Lease [7-13]). The results for thin-sheet

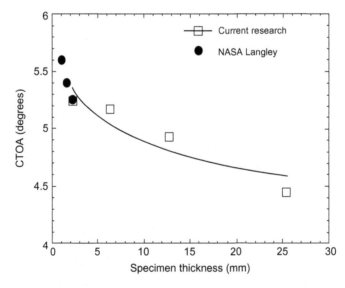

FIGURE 7.15

Thickness dependence of critical CTOA (in the constant region) for 2024-T351 aluminum alloy (adapted from Mahmoud and Lease [7-13]).

specimens in earlier NASA studies are also included. The critical CTOA values are about 5.24, 5.17, 4.92, and 4.48 degrees for the specimens 2.3, 6.35, 12.7, and 25.4 mm thick, respectively. Hence, the critical CTOA decreases with an increase in specimen thickness. Does the CTOA approach a constant value for specimens of large thickness? It appears that the answer is yes, but further experimental work needs to be done in the future.

Simulation of ductile fracture from crack initiation to extension using the CTOA criterion generally requires employment of the entire CTOA curve. Newman et al. [7-12], however, used a constant CTOA value (the CTOA in the constant region) from crack initiation to extension to simulate the load versus crack extension responses. They found that while neither the plane stress nor plane strain model produced results that matched the measured response, the three-dimensional modeling results matched the measured response by using a constant CTOA value. They attributed the phenomenon to the three-dimensional constraint effects and crack tunneling as the measured CTOA data is for the specimen surface and it is larger than that in the specimen interior.

References

[7-1] H-O. Kim, Elastic-plastic fracture analysis for small scale yielding, PhD thesis, School of Aeronautics and Astronautics, Purdue University, West Lafayette, Indiana, 1996.

[7-2] G.H. Bray, R.J. Bucci, J.R. Weh, Y. Macheret, Prediction of wide-cracked-panel toughness from small coupon tests, in: Advanced Aerospace Materials/Process Conference, Anaheim, CA, 1994.

[7-3] J.A. Begley, J.D. Landes, The J-integral as a fracture criterion, Fracture Toughness, Part II, ASTM STP 514, in: American Society for Testing and Materials, Philadelphia, 1972, pp. 1–20.

[7-4] J.W. Hutchinson, P.C. Paris, Stability analyses of J controlled crack growth, Elastic-Plastic Fracture, ASTM STP 668, in: American Society for Testing and Materials, Philadelphia, 1979, pp. 37–64.

[7-5] R.M. McMeeking, D.M. Parks, On criterion for J-dominance of crack tip fields in large scale yielding, Elastic-Plastic Fracture, ASTM STP 668, in: American Society for Testing and Materials, Philadelphia, 1979, pp. 175–194.

[7-6] C.F. Shih, M.D. German, Requirements for a one parameter characterization of crack tip fields by the HRR singularity, In. J. Fract. 17 (1981) 27–43.

[7-7] A.A. Wells, Application of fracture mechanics at and beyond general yielding, Br. Weld. J. 11 (1961) 563–570.

[7-8] D.M. Tracy, Finite element solutions for crack tip behavior in small scale yielding, ASME J. Eng. Mater. Technol. 98 (1976) 146–151.

[7-9] C.F. Shih, Relationship between the J-integral and crack opening displacement for stationary and growing cracks, J. Mech. Phys. Sol. 29 (1981) 305–326.

[7-10] J.R. Rice, Elastic-plastic crack growth, in: H.G. Hopkins, M.J. Sewell (Eds.), Mechanics of Solids, Pergamon, Oxford, UK, 1982, pp. 539–562.

[7-11] J.C. Newman Jr., U. Zerbst, Engineering fracture mechanics, Eng. Fract. Mech. 70 (2003) 367–369.

[7-12] J.C. Newman Jr., M.A. James, U. Zerbst, A review of the CTOA/CTOD fracture criterion, Eng. Fract. Mech. 70 (2003) 371–385.

[7-13] S. Mahmoud, K. Lease, The effect of specimen thickness on the experimental characterization of critical crack tip opening angle in 2024-T351 aluminum alloy, Eng. Fract. Mech. 70 (2003) 443–456.

PROBLEMS

7.1 A thin plate with a center crack is loaded with uniform tensile stresses as shown in Figure 7.16. The elastic-perfectly plastic ductile material has Young's modulus of 70 GPa and yield stress of 400 MPa. Fracture tests are conducted and failure is found to occur when the net section stress reaches the yield stress. Plot the applied stress at failure as a function of a/W. If the net-section failure is interpreted as fracture failure, find fracture toughness K_c using Irwin's plastic zone adjustment method and plot it as a function of a/W. This plot will indicate that K_c is specimen size dependent and is not suitable for characterizing this type of "ductile fracture."

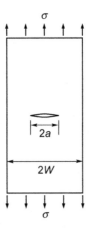

FIGURE 7.16

A center-cracked plate under tension.

7.2 Find the critical applied stress for unstable crack extension in an edge-cracked specimen (see Figure 7.17) made of a C188-T3 aluminum using the K_R curve shown in Figure 7.6. Assume that the crack length is $a = 50$ cm, the specimen

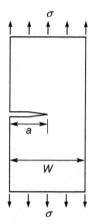

FIGURE 7.17

An edge-cracked plate under tension.

width is $W = 200$ cm, and the specimen length is much larger than the width. Use the following stress intensity factor formula for the edge-crack problem:

$$K_I = \sigma \sqrt{\pi a}(1.12 - 0.231\alpha + 10.55\alpha^2 - 21.72\alpha^3$$
$$+ 30.39\alpha^4), \alpha = a/W \leq 0.6$$

Interfacial Cracks between Two Dissimilar Solids

Engineering structures may be made of two or more dissimilar materials. Examples include structural composites, coating-substrate systems, and multilayered electronic devices. In a medium consisting of two or more materials with distinct properties, failure often initiates at the interfaces. Thus, the behavior of interface cracks is of interest. A distinct feature of a crack along the interface between two dissimilar elastic materials is that the stress field possesses an oscillatory character of the type $r^{-1/2+i\epsilon}$, where r is the radial distance from the crack tip and ϵ is a bimaterial constant. Mathematically, the oscillation behavior of the near-tip stress and displacement fields implies overlap of the upper and lower crack surfaces near the tip, which is physically inadmissible. Nevertheless, fracture criteria based on the oscillatory fields have been proposed and may be used to predict failure of interfaces because the region of the two crack faces in contact is extremely small compared with the crack length for most engineering applications.

8.1 CRACK TIP FIELDS

As in the case for homogeneous materials, the crack tip stress and displacement fields also play a central role in establishing fracture parameters for investigating interface fracture. The asymptotic eigenfunction expansion method can still be used to analyze the interface crack tip fields but the characteristic equations for determining the eigenvalue or the stress singularity index is much more complex due to the material property mismatch along the interface. In the following, Williams' solution for in-plane deformation is first presented. The solution approach is subsequently applied to the antiplane deformation case. Finally, Dundurs' parameters for describing bimaterial mismatch are introduced.

8.1.1 Asymptotic Stress and Displacement Fields

In analyzing the asymptotic crack tip stress and displacement fields, it is sufficient to consider a semi-infinite crack lying at the interface between two dissimilar homogeneous materials as shown in Figure 8.1. The upper material (material 1) has a shear modulus μ_1 and a Poisson's ratio ν_1 and the lower material (material 2) possesses

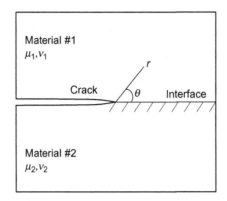

FIGURE 8.1

A crack at the interface between two dissimilar homogeneous materials.

the properties μ_2 and v_2. This problem was first analyzed by Williams [8-1] using an eigenfunction expansion method.

Let the stress functions in material 1 and material 2 be denoted by ϕ_1 and ϕ_2, respectively. These functions satisfy the following biharmonic equations:

$$\nabla^2 \nabla^2 \phi_j = 0, \quad j = 1, 2 \tag{8.1}$$

where

$$\nabla^2 = \frac{\partial^2}{\partial r^2} + \frac{1}{r}\frac{\partial}{\partial r} + \frac{1}{r^2}\frac{\partial^2}{\partial \theta^2}$$

is the Laplacian operator with (r, θ) being the polar coordinates centered at the crack tip and $\theta = 0$ along the interface, as shown in Figure 8.1. The boundary conditions of the problem consist of the crack surface ($\theta = \pm\pi$) traction-free conditions,

$$\begin{aligned}
(\sigma_{\theta\theta})_1 &= (\sigma_{r\theta})_1 = 0, \quad \theta = \pi \\
(\sigma_{\theta\theta})_2 &= (\sigma_{r\theta})_2 = 0, \quad \theta = -\pi
\end{aligned} \tag{8.2}$$

and the stress and displacement continuity conditions along the interface ($\theta = 0$),

$$\begin{aligned}
(\sigma_{\theta\theta})_1 &= (\sigma_{\theta\theta})_2, \quad \theta = 0 \\
(\sigma_{r\theta})_1 &= (\sigma_{r\theta})_2, \quad \theta = 0 \\
(u_r)_1 &= (u_r)_2, \quad \theta = 0 \\
(u_\theta)_1 &= (u_\theta)_2, \quad \theta = 0
\end{aligned} \tag{8.3}$$

A general solution to Eq. (8.1) can be obtained using the method of separation of variables by assuming

$$\phi_j = r^{\lambda+1} F_j(\theta), \quad j = 1,2 \tag{8.4}$$

where λ is the eigenvalue to be determined and $F_j(\theta)$ are the eigenfunctions. It is noted that the same value of λ is assumed for both material 1 and material 2, which is a result of continuity requirements Eq. (8.3). Substitution of Eq. (8.4) into Eq. (8.1) yields

$$\frac{d^4 F_j(\theta)}{d\theta^4} + 2(\lambda^2 + 1)\frac{d^2 F_j(\theta)}{d\theta^2} + (\lambda^2 - 1)^2 F_j(\theta) = 0, \quad j = 1,2$$

The differential equations here have the following solutions:

$$F_j(\theta) = a_j \sin(\lambda+1)\theta + b_j \cos(\lambda+1)\theta$$
$$+ c_j \sin(\lambda-1)\theta + d_j \cos(\lambda-1)\theta, \quad j = 1,2 \tag{8.5}$$

where $a_j, b_j, c_j,$ and d_j ($j = 1,2$) are unknown constants. The corresponding stresses are

$$(\sigma_{rr})_j = \frac{1}{r^2}\frac{\partial^2 \phi_j}{\partial\theta^2} + \frac{1}{r}\frac{\partial\phi_j}{\partial r} = r^{\lambda-1}\left[F_j''(\theta) + (\lambda+1)F_j(\theta)\right]$$

$$(\sigma_{\theta\theta})_j = \frac{\partial^2 \phi_j}{\partial r^2} = r^{\lambda-1}\lambda(\lambda+1)F_j(\theta) \tag{8.6}$$

$$(\sigma_{r\theta})_j = -\frac{1}{r}\frac{\partial^2 \phi_j}{\partial r\partial\theta} + \frac{1}{r^2}\frac{\partial\phi_j}{\partial\theta} = -\lambda r^{\lambda-1}F_j'(\theta)$$

and the displacement components are given by

$$(u_r)_j = \frac{1}{2\mu_j}r^\lambda\{-(\lambda+1)F_j(\theta) + (1+\kappa_j)[c_j \sin(\lambda-1)\theta$$
$$+ d_j \cos(\lambda-1)\theta]\}$$

$$(u_\theta)_j = \frac{1}{2\mu_j}r^\lambda\{-F_j'(\theta) - (1+\kappa_j)[c_j \cos(\lambda-1)\theta$$
$$- d_j \sin(\lambda-1)\theta]\} \tag{8.7}$$

where

$$\kappa_j = \begin{cases} 3 - 4\nu_j & \text{for plane strain} \\ \dfrac{3 - \nu_j}{1 + \nu_j} & \text{for plane stress} \end{cases}$$

Substitution of these stress and displacement expressions into the boundary conditions Eqs. (8.2) and (8.3) leads to the following eight simultaneous equations for the constants $a_j, b_j, c_j,$ and d_j $(j = 1, 2)$:

$$a_1 \sin(\lambda + 1)\pi + b_1 \cos(\lambda + 1)\pi + c_1 \sin(\lambda - 1)\pi + d_1 \cos(\lambda - 1)\pi = 0$$

$$-a_2 \sin(\lambda + 1)\pi + b_2 \cos(\lambda + 1)\pi - c_2 \sin(\lambda - 1)\pi + d_2 \cos(\lambda - 1)\pi = 0$$

$$a_1(\lambda + 1)\cos(\lambda + 1)\pi - b_1(\lambda + 1)\sin(\lambda + 1)\pi$$
$$+ c_1(\lambda - 1)\cos(\lambda - 1)\pi - d_1(\lambda - 1)\sin(\lambda - 1)\pi = 0$$

$$a_2(\lambda + 1)\cos(\lambda + 1)\pi + b_2(\lambda + 1)\sin(\lambda + 1)\pi$$
$$+ c_2(\lambda - 1)\cos(\lambda - 1)\pi + d_2(\lambda - 1)\sin(\lambda - 1)\pi = 0$$

$$b_1 + d_1 = b_2 + d_2$$

$$a_1(\lambda + 1) + c_1(\lambda - 1) = a_2(\lambda + 1) + c_2(\lambda - 1)$$

$$(1 + \kappa_1)c_1 = \frac{\mu_1}{\mu_2}(1 + \kappa_2)c_2 + \left(\frac{\mu_1}{\mu_2} - 1\right)[(\lambda + 1)a_2 + (\lambda - 1)c_2]$$

$$(1 + \kappa_1)d_1 = \frac{\mu_1}{\mu_2}(1 + \kappa_2)d_2 - \left(\frac{\mu_1}{\mu_2} - 1\right)(\lambda + 1)[b_2 + d_2]$$

The existence of a nontrivial solution for the four unknowns $a_j, b_j, c_j,$ and d_j requires that the determinant of the system of equations vanish. This yields the following characteristic equation to determine the eigenvalue λ:

$$\cot^2 \lambda \pi + \left[\frac{\frac{\mu_1}{\mu_2}(1 + \kappa_2) - (1 + \kappa_1) - 2\left(\frac{\mu_1}{\mu_2} - 1\right)}{\frac{\mu_1}{\mu_2}(1 + \kappa_2) + (1 + \kappa_1)} \right]^2 = 0 \qquad (8.8)$$

Except for the case of a homogeneous solid ($\mu_1 = \mu_2$ and $\nu_1 = \nu_2$), the preceding equation has only complex solutions for λ. Let

$$\lambda = \lambda_R + i\lambda_I$$

Substitution of the λ into Eq. (8.8) yields two equations for λ_R and λ_I:

$$\frac{\left(\tan^2 \lambda_R \pi + 1\right)\tanh \lambda_I \pi}{\tan^2 \lambda_R \pi + \tanh^2 \lambda_I \pi} = \pm \frac{\frac{\mu_1}{\mu_2}(1 + \kappa_2) - (1 + \kappa_1) - 2\left(\frac{\mu_1}{\mu_2} - 1\right)}{\frac{\mu_1}{\mu_2}(1 + \kappa_2) + (1 + \kappa_1)}$$

$$\frac{\tan \lambda_R \pi \left(1 - \tanh^2 \lambda_I \pi\right)}{\tan^2 \lambda_R \pi + \tanh^2 \lambda_I \pi} = 0 \tag{8.9}$$

Mathematically, these equations permit two sets of solutions:

1. It is clear that the second equation in Eq. (8.9) can be satisfied if $\tan \lambda_R \pi = 0$. Now we obtain

$$\lambda_R = n, \quad n = 0, 1, 2, \ldots$$

$$\lambda_I = \pm \frac{1}{\pi} \coth^{-1} \left[\frac{\frac{\mu_1}{\mu_2}(1+\kappa_2) - (1+\kappa_1) - 2\left(\frac{\mu_1}{\mu_2} - 1\right)}{\frac{\mu_1}{\mu_2}(1+\kappa_2) + (1+\kappa_1)} \right]$$

It is seen from the λ_R and the stress expressions in Eq. (8.6) that only one value of λ_R corresponding to $n = 0$ will produce stress singularity at the crack tip. With this value, the stresses in Eq. (8.6) near the crack tip vary with r as

$$\sigma_{rr}, \sigma_{\theta\theta}, \sigma_{r\theta} \sim \frac{1}{r} \binom{\sin}{\cos} (\lambda_I \ln r)$$

It can be shown that these stresses lead to infinite strain energy in any small region containing the crack tip ($r \to 0$), which is physically unrealistic.

2. The second equation in Eq. (8.9) can also be satisfied if $\tan \lambda_R \pi = \infty$. We then obtain the second set of the eigenvalue:

$$\lambda_R = \frac{2n-1}{2}, \quad n = 1, 2, 3, \ldots$$

$$\lambda_I = \pm \frac{1}{\pi} \tanh^{-1} \left[\frac{\frac{\mu_1}{\mu_2}(1+\kappa_2) - (1+\kappa_1) - 2\left(\frac{\mu_1}{\mu_2} - 1\right)}{\frac{\mu_1}{\mu_2}(1+\kappa_2) + (1+\kappa_1)} \right]$$

Using the relation

$$\tanh^{-1} \frac{x}{b} = \frac{1}{2} \ln \frac{b+x}{b-x}$$

we can express the eigenvalues in the form

$$\lambda = \left(n - \frac{1}{2}\right) \pm i\epsilon, \quad n = 1, 2, \ldots \tag{8.10}$$

where

$$\epsilon = \frac{1}{2\pi} \ln \left[\left(\frac{\kappa_1}{\mu_1} + \frac{1}{\mu_2}\right) \bigg/ \left(\frac{\kappa_2}{\mu_2} + \frac{1}{\mu_1}\right) \right] \tag{8.11}$$

It can be seen from Eqs. (8.6) and (8.10) that only the eigenvalue for $n = 1$ produces singular stresses at the crack tip. The $\pm$ sign in Eq. (8.10) is not essential because ϵ changes sign if material 1 and 2 are switched. The eigenvalue is thus chosen as $\lambda = 1/2 + i\epsilon$ for the crack tip singular stresses, which vary with r as

$$\sigma_{rr}, \sigma_{\theta\theta}, \sigma_{r\theta} \sim r^{-1/2+i\epsilon}$$

and the crack tip displacements have the asymptotic form as $r \to 0$:

$$u_r, u_\theta \sim r^{1/2+i\epsilon}$$

Because

$$r^{i\epsilon} = e^{i\epsilon \ln r} = \cos(\epsilon \ln r) + i\sin(\epsilon \ln r)$$

the crack tip stresses and displacements can also be expressed as

$$\sigma_{rr}, \sigma_{\theta\theta}, \sigma_{r\theta} \sim r^{-1/2}[\cos(\epsilon \ln r) + i\sin(\epsilon \ln r)] \tag{8.12}$$

$$u_r, u_\theta \sim r^{1/2}[\cos(\epsilon \ln r) + i\sin(\epsilon \ln r)] \tag{8.13}$$

A few comments on the stress and displacement fields Eqs. (8.12) and (8.13) are in order:

1. The stress singularity of the type $r^{-1/2+i\epsilon}$ and the displacements' dependence on r according to $r^{1/2+i\epsilon}$ near the crack tip implies that the stresses and the displacements exhibit an oscillating character as they change sign indefinitely when r approaches zero.

2. The oscillatory nature of the displacement field implies interpenetration of two crack faces, which is physically inadmissible. In other words, the crack faces must come in contact near the tip. This will be discussed in Sections 8.3 and 8.7.

3. In the special case of $\epsilon = 0$, the oscillatory terms disappear, and the near-tip stress and displacement fields have exactly the same forms as those for homogeneous materials. ϵ is sometimes called the oscillation index.

8.1.2 Mode III Case

An interface crack in a Mode III field is a special case wherein no stress oscillation occurs as opposed to the general in-plane modes cases. Consider the bimaterial crack problem of Figure 8.1 with the medium subjected to anti-plane shear loading. The only nonvanishing displacements are $w_j = w_j(r,\theta)$, which satisfy the harmonic equations

$$\nabla^2 w_j = \frac{\partial^2 w_j}{\partial r^2} + \frac{1}{r}\frac{\partial w_j}{\partial r} + \frac{1}{r^2}\frac{\partial^2 w_j}{\partial \theta^2} = 0, \quad j = 1,2 \tag{8.14}$$

The nonvanishing stress components are given by

$$(\sigma_{rz})_j = \mu_j \frac{\partial w_j}{\partial r}, \quad (\sigma_{\theta z})_j = \mu_j \frac{1}{r} \frac{\partial w_j}{\partial \theta} \tag{8.15}$$

The boundary conditions for the Mode III crack problem consist of the traction-free conditions on the crack surfaces,

$$\begin{aligned} (\sigma_{\theta z})_1 = 0, \quad \theta = \pi \\ (\sigma_{\theta z})_2 = 0, \quad \theta = -\pi \end{aligned} \tag{8.16}$$

and the continuity of stresses and displacements along the interface,

$$\begin{aligned} (\sigma_{\theta z})_1 = (\sigma_{\theta z})_2, \quad \theta = 0 \\ w_1 = w_2, \quad \theta = 0 \end{aligned} \tag{8.17}$$

Following the treatment for the in-plane crack problem described before, the displacements and stresses near the crack tip may be expressed in the following forms after the governing Eq. (8.14) is satisfied:

$$w_j(r,\theta) = r^\lambda [A_j \sin(\lambda\theta) + B_j \cos(\lambda\theta)]$$

$$(\sigma_{rz})_j = \mu_j \lambda r^{\lambda-1} [A_j \sin(\lambda\theta) + B_j \cos(\lambda\theta)]$$
$$(\sigma_{\theta z})_j = \mu_j \lambda r^{\lambda-1} [A_j \cos(\lambda\theta) - B_j \sin(\lambda\theta)]$$

Use of the boundary conditions Eqs. (8.16) and (8.17) leads to the following equations to determine the unknown constants A_j and B_j $(j = 1, 2)$:

$$\mu_1 A_1 = \mu_2 A_2$$
$$B_1 = B_2$$

$$A_1 \cos(\lambda\pi) - B_1 \sin(\lambda\pi) = 0$$
$$\mu_1 A_1 \cos(\lambda\pi) + \mu_2 B_1 \sin(\lambda\pi) = 0$$

A nontrivial solution of A_j and B_j exists if

$$\begin{vmatrix} \cos(\lambda\pi) & -\sin(\lambda\pi) \\ \mu_1 \cos(\lambda\pi) & \mu_2 \sin(\lambda\pi) \end{vmatrix} = 0$$

or

$$\sin(2\lambda\pi) = 0$$

Thus, the eigenvalue λ can be obtained as

$$\lambda = \lambda^{(n)} = \frac{n}{2}, \quad n = 0, \pm 1, \pm 2$$

Since $n = -1, -2, \ldots$ lead to unbounded displacement at the crack tip, they are excluded in the solution. In addition, $\lambda = 0$ is neglected since it corresponds to a rigid body motion.

Consider the class of solutions for which the displacement w is antisymmetric with respect to the crack plane (x-axis). Thus,

$$B_1 = B_2 = 0$$

and

$$\lambda^{(n)} = \frac{1}{2} + n, \quad n = 0, 1, 2, 3, \ldots \tag{8.18}$$

The displacement solution is given by

$$w_1(r, \theta) = \sum_{n=0}^{\infty} r^{\frac{1}{2}+n} A^{(n)} \sin\left[\left(\frac{1}{2}+n\right)\theta\right] \tag{8.19}$$

$$w_2(r, \theta) = \frac{\mu_1}{\mu_2} w_1(r, \theta)$$

and the corresponding stresses are

$$(\sigma_{rz})_1 = (\sigma_{rz})_2 = \mu_1 \sum_{n=0}^{\infty} \left(n + \frac{1}{2}\right) r^{n-\frac{1}{2}} A^{(n)} \sin\left[\left(n + \frac{1}{2}\right)\theta\right]$$

$$(\sigma_{\theta z})_1 = (\sigma_{\theta z})_2 = \mu_1 \sum_{n=0}^{\infty} \left(n + \frac{1}{2}\right) r^{n-\frac{1}{2}} A^{(n)} \cos\left[\left(n + \frac{1}{2}\right)\theta\right] \tag{8.20}$$

It can be seen from these stress expressions that $n = 0$ yields the inverse square root ($1/\sqrt{r}$) singularity and the crack tip singular stress field can be expressed as

$$(\sigma_{rz})_1 = (\sigma_{rz})_2 = \frac{1}{2} \mu_1 A^{(0)} r^{-1/2} \sin\frac{\theta}{2}$$

$$(\sigma_{\theta z})_1 = (\sigma_{\theta z})_2 = \frac{1}{2} \mu_1 A^{(0)} r^{-1/2} \cos\frac{\theta}{2} \tag{8.21}$$

It is clear that no oscillatory behavior is seen in the Mode III stress field.

8.1.3 **Dundurs' Parameters**

Dundurs [8-2] introduced two bimaterial parameters called α and β to describe the material property mismatch:

$$\alpha = \frac{\mu_1(\kappa_2+1)-\mu_2(\kappa_1+1)}{\mu_1(\kappa_2+1)+\mu_2(\kappa_1+1)}$$

$$\beta = \frac{\mu_1(\kappa_2-1)-\mu_2(\kappa_1-1)}{\mu_1(\kappa_2+1)+\mu_2(\kappa_1+1)}$$

(8.22)

Clearly, $\alpha = \beta = 0$ if no material property mismatch occurs. For general bimaterial systems with positive Poisson's ratios, $|\alpha| \leq 1$. The equality takes place when $\mu_2/\mu_1 \to \infty$ or $\mu_2/\mu_1 \to 0$. Similarly, $|\beta| \leq 0.5$ and the equality occurs when $\mu_2/\mu_1 \to \infty$ and $\nu_1 = 0$ or $\mu_2/\mu_1 \to 0$ and $\nu_2 = 0$.

With the Dundurs parameter β, the oscillation index ϵ in the crack tip field can be expressed as

$$\epsilon = \frac{1}{2\pi}\ln\frac{1-\beta}{1+\beta}$$

(8.23)

8.2 **COMPLEX FUNCTION METHOD AND STRESS INTENSITY FACTORS**

The complex variables method of Kosolov-Muskhelishvili for homogeneous materials may still be used to obtain stresses and stress intensity factor solutions for interface cracks, although the mathematical manipulations are more complicated. The complex potential representations of stresses and displacements, Eqs. (3.23) through (3.25) in Chapter 3, now have the forms

$$(\sigma_{xx})_1 + (\sigma_{yy})_1 = 4\,\mathrm{Re}\{\psi_1'(z)\}$$

$$(\sigma_{yy})_1 - (\sigma_{xx})_1 + 2i\,(\sigma_{xy})_1 = 2\{\bar{z}\psi_1''(z)+\chi_1''(z)\}$$

(8.24)

$$2\mu_1\left[(u_x)_1 + i\,(u_y)_1\right] = \kappa_1\psi_1(z) - z\overline{\psi_1'(z)} - \overline{\chi_1'(z)}$$

for material 1 (see Figure 8.1), and

$$(\sigma_{xx})_2 + (\sigma_{yy})_2 = 4\,\mathrm{Re}\{\psi_2'(z)\}$$

$$(\sigma_{yy})_2 - (\sigma_{xx})_2 + 2i\,(\sigma_{xy})_2 = 2\{\bar{z}\psi_2''(z)+\chi_2''(z)\}$$

(8.25)

$$2\mu_2\left[(u_x)_2 + i\,(u_y)_2\right] = \kappa_2\psi_2(z) - z\overline{\psi_2'(z)} - \overline{\chi_2'(z)}$$

for material 2 (see Figure 8.1).

It follows from the preceding expressions and the crack tip stress field Eq. (8.12) that near the crack tip,

$$\psi_1'(z) \sim (z-L)^{-1/2-i\epsilon}, \quad z \to L$$

where $z = L$ represents the crack tip. Rice and Sih [8-3] thus defined a complex stress intensity factor by

$$k_1 - ik_2 = 2\sqrt{2}e^{\pi\epsilon} \lim_{z \to L} (z-L)^{1/2+i\epsilon} \psi_1'(z) \qquad (8.26)$$

8.2.1 Stress Intensity Factor Solutions for Two Typical Crack Problems

Rice and Sih [8-3] solved two typical interface crack problems as shown in Figures 8.2 and 8.3. A detailed description for the solution procedure is not provided here. For a crack of length $2a$ at the interface between two semi-infinite media subjected to remote uniform tension σ_{yy}^{∞} and shear σ_{xy}^{∞} as shown in Figure 8.2, Rice and Sih [8-3] gave the following solutions of the complex potentials:

$$\psi_1'(z) = g(z)F(z) + A$$

$$\chi_1''(z) = e^{2\pi\epsilon}\overline{g}(z)\overline{F}(z) + \left[\frac{a^2 + 2i\epsilon az}{z^2 - a^2}g(z) - zg'(z)\right]F(z) - (A + \overline{A})$$

where $F(z)$ is

$$F(z) = \frac{1}{\sqrt{z^2 - a^2}}\left(\frac{z+a}{z-a}\right)^{i\epsilon}$$

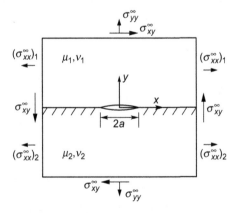

FIGURE 8.2

A crack of length $2a$ at the interface between two semi-infinite dissimilar media subjected to remote uniform loading.

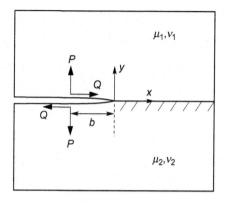

FIGURE 8.3

A semi-infinite interface crack subjected to crack surface concentrated loads.

$g(z)$ is given by

$$g(z) = \frac{\sigma_{yy}^{\infty} - i\sigma_{xy}^{\infty}}{1 + e^{2\pi\epsilon}}(z - 2i\epsilon a)$$

and A is a constant

$$A = A_1 + iA_2$$

$$A_1 = \frac{(\sigma_{xx}^{\infty})_1 + \sigma_{yy}^{\infty}}{4} - \frac{\sigma_{yy}^{\infty}}{1 + e^{2\pi\epsilon}}$$

$$A_2 = \frac{\sigma_{xy}^{\infty}}{1 + e^{2\pi\epsilon}} + \frac{2\mu_1\omega_1^{\infty}}{1 + \kappa_1} = \frac{1}{q}\left(\frac{e^{2\pi\epsilon}\sigma_{xy}^{\infty}}{1 + e^{2\pi\epsilon}} + \frac{2\mu_2\omega_2^{\infty}}{1 + \kappa_2}\right)$$

with ω_1^{∞} and ω_2^{∞} representing the rotations at infinity in the upper and lower half planes, respectively, and q being given by

$$q = \frac{\mu_2(\kappa_1 + 1)}{\mu_1(\kappa_2 + 1)}$$

With these complex potentials, the stress intensity factors can be calculated from Eq. (8.26) (now $L = a$) as

$$k_1 = \{\sigma\left[\cos(\epsilon\ln 2a) + 2\epsilon\sin(\epsilon\ln 2a)\right] + \tau\left[\sin(\epsilon\ln 2a) - 2\epsilon\cos(\epsilon\ln 2a)\right]\}$$
$$\times \sqrt{a}/\cosh(\pi\epsilon)$$

$$\text{(8.27)}$$

$$k_2 = \{\tau\left[\cos(\epsilon\ln 2a) + 2\epsilon\sin(\epsilon\ln 2a)\right] - \sigma\left[\sin(\epsilon\ln 2a) - 2\epsilon\cos(\epsilon\ln 2a)\right]\}$$
$$\times \sqrt{a}/\cosh(\pi\epsilon)$$

where $\sigma = \sigma_{yy}^{\infty}$ and $\tau = \sigma_{xy}^{\infty}$. Note that each stress intensity factor is induced by both tension and shear loads. The asymptotic stresses along the interface ahead of the crack tip can be expressed as

$$\left(\sigma_{yy} + i\sigma_{xy}\right)_{\theta=0} = \frac{k_1 - ik_2}{\sqrt{2r}} r^{i\epsilon} \cosh(\pi\epsilon) \tag{8.28}$$

and the near-tip relative displacements along the crack surfaces as

$$\left(u_y + iu_x\right)_{\theta=\pi} - \left(u_y + iu_x\right)_{\theta=-\pi} = \left(\frac{\kappa_1 + 1}{\mu_1} + \frac{\kappa_2 + 1}{\mu_2}\right) \frac{(k_1 - ik_2)\sqrt{2r}}{4(1 + 2i\epsilon)} r^{i\epsilon} \cosh(\pi\epsilon) \tag{8.29}$$

For a semi-infinite interface crack between two semi-infinite media subjected to concentrated forces P (tensile) and Q (shearing) at a distance b from the crack tip, as shown in Figure 8.3, Rice and Sih [8-3] gave the following solutions of the complex potentials:

$$\psi'_1(z) = z^{-1/2-i\epsilon} f(z)$$

$$\chi''_1(z) = e^{2\pi\epsilon} z^{-1/2+i\epsilon} \overline{f}(z) - z^{-1/2-i\epsilon} \left[(1/2 - i\epsilon)f(z) + zf'(z)\right]$$

where

$$f(z) = \frac{P - iQ}{2\pi e^{\pi\epsilon}} \frac{b^{1/2+i\epsilon}}{z+b}$$

With these complex potentials, the stress intensity factors are calculated as follows:

$$k_1 = \frac{1}{\pi}\sqrt{\frac{2}{b}}[P\cos(\epsilon \ln b) + Q\sin(\epsilon \ln b)]$$

$$k_2 = \frac{1}{\pi}\sqrt{\frac{2}{b}}[Q\cos(\epsilon \ln b) - P\sin(\epsilon \ln b)] \tag{8.30}$$

Note that in this example, the crack tip is at $z = L = 0$ in Eq. (8.26).

8.2.2 Further Comments on the Stress Intensity Factor Definitions

It is seen from Eq. (8.27) that the stress intensity factors depend on the measuring unit of the crack length because the $\ln(a)$ term is involved. The stress intensity factor thus is not uniquely determined. Furthermore, Mode I and Mode II deformations are coupled together even if the external loads and specimen geometry possess symmetry about the crack line. Thus, a pure Mode I or Mode II stress intensity factor can not be clearly defined. Several definitions of stress intensity factors have been proposed depending on the treatment of the coupling term $r^{i\epsilon}$ in the stress field Eq. (8.28) (e.g., in [8-3] through [8-6]). Most of these definitions differ by either a phase factor or a constant.

Hutchinson et al. [8-4] introduced a complex stress intensity factor K_h in such a way that, along the interface ahead of the crack tip, stresses are given by

$$(\sigma_{yy} + i\sigma_{xy})_{\theta=0} = \frac{K_h}{\sqrt{2\pi r}} r^{i\epsilon} \tag{8.31}$$

Comparing this expression with Eq. (8.28), we see that K_h is related to the stress intensity factor $K_r = k_1 - ik_2$ of Rice and Sih [8-3] by

$$K_r = K_h / \left[\sqrt{\pi} \cosh(\pi\epsilon)\right] \tag{8.32}$$

For the interface crack of length $2a$ in an infinite bimaterial subjected to remotely uniform stresses σ_{yy}^∞ and σ_{xy}^∞ as shown in Figure 8.2, the complex stress intensity factor using the definition of Hutchinson et al. [8-4] is

$$K_h = (\sigma_{yy}^\infty + i\sigma_{xy}^\infty)(1 + 2i\epsilon)(2a)^{-i\epsilon} \sqrt{\pi a} \tag{8.33}$$

Similar to K_r, K_h in Eq. (8.33) is also a function of the measuring unit of the crack length. Both K_h and K_r contain a crack length–related phase term $(2a)^{-i\epsilon}$ giving rise to a complex dimension for the stress intensity factor.

Malysev and Salganik [8-5] introduced a stress intensity factor by expressing the crack tip stresses as

$$(\sigma_{yy} + i\sigma_{xy})_{\theta=0} = \frac{K_m}{\sqrt{2\pi r}} \left(\frac{r}{2a}\right)^{i\epsilon} \tag{8.34}$$

For the interface crack problem shown in Figure 8.2

$$K_m = (\sigma_{yy}^\infty + i\sigma_{xy}^\infty)(1 + 2i\epsilon)\sqrt{\pi a} \tag{8.35}$$

Sun and Jih [8-6] introduced a similar stress intensity factor $K = K_1 + iK_2$ with the crack tip stress field expressed by

$$(\sigma_{yy} + i\sigma_{xy})_{\theta=0} = \frac{K}{\sqrt{2\pi r}} \left(\frac{r}{2a}\right)^{i\epsilon} \cosh(\pi\epsilon) \tag{8.36}$$

The relationship between K and K_h is

$$K_h = K(2a)^{-i\epsilon} \cosh(\pi\epsilon) \tag{8.37}$$

Both K_m and K remove the ambiguity of dimension by nondimensionalizing r in $r^{i\epsilon}$ with $2a$ in the near-tip stress field. As a result, both have the same dimension as that of the classical stress intensity factor. All the previous definitions of stress intensity factor are acceptable as far as the near-tip state is concerned. With the definition of K by Sun and Jih [8-6], the stress intensity factors for the finite crack problem shown in Figure 8.2 become

$$K_1 = \sqrt{\pi a} \left(\sigma_{yy}^{\infty} - 2\epsilon \sigma_{xy}^{\infty} \right) / \cosh(\pi \epsilon)$$

$$K_2 = \sqrt{\pi a} \left(\sigma_{xy}^{\infty} + 2\epsilon \sigma_{yy}^{\infty} \right) / \cosh(\pi \epsilon)$$

(8.38)

Using the stress intensity factor definition in Eq. (8.36), the near-tip relative displacement field Eq. (8.29) along the crack surfaces becomes

$$\left(u_y + i u_x \right)_{\theta = \pi} - \left(u_y + i u_x \right)_{\theta = -\pi}$$

$$= \left(\frac{\kappa_1 + 1}{\mu_1} + \frac{\kappa_2 + 1}{\mu_2} \right) \frac{K_1 + i K_2}{4(1 + 2i\epsilon)} \sqrt{\frac{2r}{\pi}} \left(\frac{r}{2a} \right)^{i\epsilon}$$

(8.39)

Sun and Jih [8-6] have given the stress and displacement fields around an interface crack tip as follows:

$$(\sigma_{xx})_j = \frac{K_1}{2\sqrt{2\pi r}} \left[\omega_j f_{xx}^I - \frac{1}{\omega_j} \cos \left(\theta - \overline{\Theta} \right) \right]$$

$$- \frac{K_2}{2\sqrt{2\pi r}} \left[\omega_j f_{xx}^{II} + \frac{1}{\omega_j} \sin \left(\theta - \overline{\Theta} \right) \right]$$

$$(\sigma_{yy})_j = \frac{K_1}{2\sqrt{2\pi r}} \left[\omega_j f_{yy}^I + \frac{1}{\omega_j} \cos \left(\theta - \overline{\Theta} \right) \right]$$

$$- \frac{K_2}{2\sqrt{2\pi r}} \left[\omega_j f_{yy}^{II} - \frac{1}{\omega_j} \sin \left(\theta - \overline{\Theta} \right) \right]$$

(8.40)

$$(\sigma_{xy})_j = \frac{K_1}{2\sqrt{2\pi r}} \left[\omega_j f_{xy}^I - \frac{1}{\omega_j} \sin \left(\theta - \overline{\Theta} \right) \right]$$

$$- \frac{K_2}{2\sqrt{2\pi r}} \left[\omega_j f_{xy}^{II} - \frac{1}{\omega_j} \cos \left(\theta - \overline{\Theta} \right) \right]$$

and

$$(u_x)_j = \frac{K_1 \sqrt{2\pi r}}{4\pi \mu_j} \left[\kappa_j \omega_j h_{11} - \frac{1}{\omega_j} h_{12} + \omega_j h_{13} \right]$$

$$+ \frac{K_2 \sqrt{2\pi r}}{4\pi \mu_j} \left[\kappa_j \omega_j h_{21} - \frac{1}{\omega_j} h_{22} + \omega_j h_{23} \right]$$

(8.41)

$$(u_y)_j = \frac{K_1 \sqrt{2\pi r}}{4\pi \mu_j} \left[\kappa_j \omega_j h_{21} - \frac{1}{\omega_j} h_{22} - \omega_j h_{23} \right]$$

$$+ \frac{K_2 \sqrt{2\pi r}}{4\pi \mu_j} \left[-\kappa_j \omega_j h_{11} + \frac{1}{\omega_j} h_{12} + \omega_j h_{13} \right]$$

where $j = 1, 2$ represent the quantities in material 1 and 2, respectively, and

$$\overline{\Theta} = \epsilon \ln\left(\frac{r}{2a}\right) + \frac{\theta}{2}$$

$$\omega_1 = \exp[-\epsilon(\pi - \theta)]$$

$$\omega_2 = \exp[\epsilon(\pi + \theta)]$$

$$f_{xx}^I = 3\cos\overline{\Theta} + 2\epsilon\sin\theta\cos(\theta + \overline{\Theta}) - \sin\theta\sin(\theta + \overline{\Theta})$$

$$f_{xx}^{II} = 3\sin\overline{\Theta} + 2\epsilon\sin\theta\sin(\theta + \overline{\Theta}) + \sin\theta\cos(\theta + \overline{\Theta})$$

$$f_{yy}^I = \cos\overline{\Theta} - 2\epsilon\sin\theta\cos(\theta + \overline{\Theta}) + \sin\theta\sin(\theta + \overline{\Theta})$$

$$f_{yy}^{II} = \sin\overline{\Theta} - 2\epsilon\sin\theta\sin(\theta + \overline{\Theta}) - \sin\theta\cos(\theta + \overline{\Theta})$$

$$f_{xy}^I = \sin\overline{\Theta} + 2\epsilon\sin\theta\sin(\theta + \overline{\Theta}) + \sin\theta\cos(\theta + \overline{\Theta})$$

$$f_{xy}^{II} = -\cos\overline{\Theta} - 2\epsilon\sin\theta\cos(\theta + \overline{\Theta}) + \sin\theta\sin(\theta + \overline{\Theta})$$

$$h_{11} = \frac{1}{1 + 4\epsilon^2}\left[\cos(\theta - \overline{\Theta}) - 2\epsilon\sin(\theta - \overline{\Theta})\right]$$

$$h_{12} = \frac{1}{1 + 4\epsilon^2}\left[\cos\overline{\Theta} - 2\epsilon\sin\overline{\Theta}\right]$$

$$h_{13} = \sin\theta\sin\overline{\Theta}$$

$$h_{21} = \frac{1}{1 + 4\epsilon^2}\left[\sin(\theta - \overline{\Theta}) + 2\epsilon\cos(\theta - \overline{\Theta})\right]$$

$$h_{22} = \frac{1}{1 + 4\epsilon^2}\left[-\sin\overline{\Theta} + 2\epsilon\cos\overline{\Theta}\right]$$

$$h_{23} = \sin\theta\cos\overline{\Theta}$$

8.3 CRACK SURFACE CONTACT ZONE AND STRESS OSCILLATION ZONE

In this section, the crack surface contact zone and the stress oscillation zone near an interface crack tip are analyzed using the oscillatory crack tip fields introduced in Sections 8.2 and 8.3. More precise analysis of crack surface contact considering crack surface friction under shear dominated loading will be provided in Section 8.7.

8.3.1 Crack Surface Contact Zone

The crack surface displacements Eq. (8.39) exhibit an oscillating character, that is, they change sign indefinitely when r approaches zero. This oscillatory nature means that the upper and lower crack surfaces will overlap near the crack tip, which is physically unrealistic. The two crack surfaces are actually in contact near the tip. In order for the stress intensity factors based on the oscillatory crack tip fields Eq. (8.36) to predict interface fracture, the size of this contact zone must be much smaller than the crack length or any other in-plane dimensions of the cracked body (small-scale contact).

Although precise evaluation of the contact zone size requires consideration of the crack surface contact in the analysis of displacement field (see [8-7] through [8-8]), which will be discussed in Section 8.7, the crack tip oscillatory displacement solution Eq. (8.39) provides a reasonable estimate for the contact zone size if small-scale contact prevails. The near-tip relative displacement along the crack surfaces can be obtained from Eq. (8.39) as follows:

$$\Delta u_y = u_y(r, \pi) - u_y(r, -\pi)$$
$$= \frac{\sqrt{2r}}{4(1 + 4\epsilon^2)\sqrt{\pi}} \left(\frac{\kappa_1 + 1}{\mu_1} + \frac{\kappa_2 + 1}{\mu_2} \right) (K_1 H_1 - K_2 H_2)$$

$$\Delta u_x = u_x(r, \pi) - u_x(r, -\pi)$$
$$= \frac{\sqrt{2r}}{4(1 + 4\epsilon^2)\sqrt{\pi}} \left(\frac{\kappa_1 + 1}{\mu_1} + \frac{\kappa_2 + 1}{\mu_2} \right) (K_1 H_2 + K_2 H_1)$$

(8.42)

where

$$H_1 = \left[\cos\left(\epsilon \ln\left(\frac{r}{2a} \right) \right) + 2\epsilon \sin\left(\epsilon \ln\left(\frac{r}{2a} \right) \right) \right]$$
$$H_2 = \left[\sin\left(\epsilon \ln\left(\frac{r}{2a} \right) \right) - 2\epsilon \cos\left(\epsilon \ln\left(\frac{r}{2a} \right) \right) \right]$$

(8.43)

The contact zone size may be estimated by adopting r_c, the largest r at which the relative normal displacement given by Eq. (8.42) vanishes, that is,

$$K_1 H_1 - K_2 H_2 = 0$$

Substitution of Eq. (8.43) into this equation yields (Sun and Qian [8-9])

$$r_c = 2a \exp\left[\frac{1}{\epsilon} \left(\tan^{-1}\left(\frac{K_1 + 2\epsilon \cdot K_2}{K_2 - 2\epsilon \cdot K_1} \right) - \pi \right) \right]$$

(8.44)

Equation (8.44) represents the contact zone size for general crack problems. For an infinite bimaterial medium with a crack subjected to remote loading shown in Figure 8.2, Eq. (8.44) becomes

$$r_c = 2a \exp\left[-\frac{1}{\epsilon} \left(\frac{\pi}{2} + \psi \right) \right]$$

(8.45)

where ψ is the loading phase angle defined by

$$\sigma_{yy}^{\infty} + i\sigma_{xy}^{\infty} = Te^{i\psi}$$

Redefining the phase angle ψ as

$$\tan\psi = \frac{K_2}{K_1}$$

Eq. (8.44) can be rewritten in the following form:

$$\frac{r_c}{2a} = \exp\left[\frac{1}{\epsilon}\left(\tan^{-1}\left(\frac{1+2\epsilon\tan\psi}{\tan\psi - 2\epsilon}\right) - \pi\right)\right] \tag{8.46}$$

For the small-scale contact conditions to prevail, Rice [8-10] suggested that $r_c/(2a) \le 0.01$. Using Eq. (8.46), this condition can be expressed as

$$\frac{1+2\epsilon\tan\psi}{\tan\psi - 2\epsilon} \le \tan(\epsilon\ln 0.01)$$

8.3.2 Stress Oscillation Zone

The oscillatory nature of the crack tip stress field Eq. (8.36) implies that the normal stress σ_{yy} may become negative near the crack tip even if the cracked bimaterial is subjected to tension loading. To study the size of the stress oscillation zone, we consider the near-tip stress fields for a two-dimensional infinite medium with a center crack of size $2a$ subjected to remotely uniform tensile stress σ_{yy}^{∞} shown in Figure 8.2 (with $\sigma_{xy}^{\infty} = 0$).

The asymptotic crack tip stresses along the interface ahead of the crack tip for this problem can be obtained by substituting Eq. (8.38) into Eq. (8.36):

$$\sigma_{yy} = \frac{\sigma_{yy}^{\infty}}{\sqrt{2r/a}}\left[\cos\left(\epsilon\ln\frac{r}{2a}\right) - 2\epsilon\sin\left(\epsilon\ln\frac{r}{2a}\right)\right]$$

$$\sigma_{xy} = \frac{\sigma_{yy}^{\infty}}{\sqrt{2r/a}}\left[\sin\left(\epsilon\ln\frac{r}{2a}\right) + 2\epsilon\cos\left(\epsilon\ln\frac{r}{2a}\right)\right]$$

Let r_o and r_o^* denote the sizes of the oscillation zones of the normal stress and the shear stress, respectively. Without loss of generality, we assume $\epsilon > 0$. The size r_o for the normal stress σ_{yy} can be determined by finding the largest value of x at which $d\sigma_{yy}/dx = 0$ with the result

$$r_o = 2a\exp\left(-\frac{\pi}{2\epsilon}\right) \tag{8.47}$$

It is noted that r_o is the same as the crack surface contact zone size given in Eq. (8.45) (now $\psi = 0$). Similarly, by letting $d\sigma_{xy}/dx = 0$, we can find the oscillation zone size

r_o^* for the shear stress as

$$r_o^* = 2a \exp\left(-\frac{\pi}{\epsilon}\right) \tag{8.48}$$

These two equations indicate that the size of the oscillation zone for the shear stress is much smaller than that for the normal stress and the oscillation zone size depends on the parameter ϵ, which is uniquely expressed by Dundurs' parameter β in Eq. (8.23).

For the largest mismatch case of $\beta = 0.5$ ($\epsilon = 0.1748$), the size of the normal stress oscillation zone is evaluated as $r_o/a = 2.51 \times 10^{-4}$ and the shear stress oscillation zone size is $r_o^*/a = 3.15 \times 10^{-8}$. For $\beta = 0.2$, these lengths are evaluated as $r_o/a = 5.34 \times 10^{-11}$ and $r_o^* = 1.44 \times 10^{-21}$, respectively. Hence, except for extreme mismatch cases, the oscillation zone is very small as compared with the crack size.

Gautsen and Dundurs [8-11] obtained the stresses along the interface for the crack problem shown in Figure 8.2 (remote tensile load only) by considering the crack face contact in the near-tip region. The log-log plot of the normal and shear stresses ahead of the crack tip from the contact model [8-11] and the oscillatory model [8-3] are shown in Figures 8.4 and 8.5 for $\beta = 0.2$ and $\beta = 0.5$, respectively, where $\beta = 0.5$ stands for the largest mismatch of elastic constants.

It is clear from the figures that solutions from both models agree extremely well beyond $x/a = 10^{-4}$, except for the normal stress associated with $\beta = 0.5$. For this case, the normal stresses from the two models agree up to $x/a = r_o \approx 2.51 \times 10^{-4}$. This value turns out to be the size of the oscillation (or overlap) zone for the crack-tip normal stress. Because the stresses of the two models agree very well except in the small contact zone near the crack tip, the oscillatory crack tip fields may still be used as the basis for predicting interface fracture.

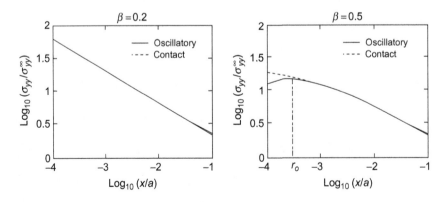

FIGURE 8.4

Normal stress distributions along the interface ahead of the crack with and without consideration of crack surface contact (adapted from Sun and Qian [8-9]).

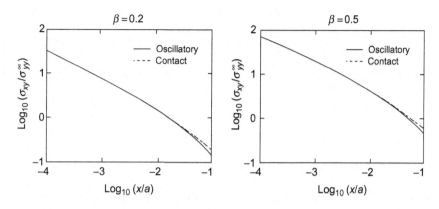

FIGURE 8.5

Shear stress distributions along the interface ahead of the crack with and without consideration of crack surface contact (adapted from Sun and Qian [8-9]).

8.4 ENERGY RELEASE RATE

Both Mode I and Mode II are, in general, present in interface cracks. Moreover, the energy release rates for the individual modes do not exist because of the oscillatory nature of the crack tip stress and displacement fields. Since the total energy release rate exists, the fracture criterion for interface cracks is usually given in terms of a critical value of the total energy release rate accompanied by a mode mixity $\tan^{-1}$ (K_{II}/K_1). Thus, in addition to the total energy release rate, both stress intensity factors must be calculated for fracture prediction.

8.4.1 Energy Release Rate

Due to the complication related to the oscillatory stress and displacement fields, the energy balance approach is more convenient for inteface fracture. Unlike the case for homogeneous materials, the Mode I and Mode II energy release rates do not exist for general bimaterial interface cracks (Sun and Jih [8-6]). Using the crack closure integral technique of Irwin [8-12], the Mode I and Mode II energy release rates may be defined as

$$G_1 = \lim_{\Delta a \to 0} \frac{1}{2\Delta a} \int_0^{\Delta a} \sigma_{yy}(x,0) \Delta u_y(\Delta a - x, \pi) dx$$

$$G_2 = \lim_{\Delta a \to 0} \frac{1}{2\Delta a} \int_0^{\Delta a} \sigma_{xy}(x,0) \Delta u_x(\Delta a - x, \pi) dx$$

(8.49)

where $\sigma_{xy}(x,0)$ and $\sigma_{yy}(x,0)$ are the stress fields when the crack tip is at $x = 0$, and $\Delta u_x(\Delta a - x, \pi)$ and $\Delta u_y(\Delta a - x, \pi)$ are the relative displacements of the crack faces when the crack tip is at $x = \Delta a$.

The normal and shear stresses at the interface ahead of the crack tip ($x = 0$) are obtained from Eq. (8.36) as

$$\sigma_{yy}(x,0) = \frac{\cosh(\pi\epsilon)}{\sqrt{2\pi x}}\left[K_1(a)\cos\left(\epsilon\ln\frac{x}{2a}\right) - K_2(a)\sin\left(\epsilon\ln\frac{x}{2a}\right)\right]$$

$$\sigma_{xy}(x,0) = \frac{\cosh(\pi\epsilon)}{\sqrt{2\pi x}}\left[K_1(a)\sin\left(\epsilon\ln\frac{x}{2a}\right) + K_2(a)\cos\left(\epsilon\ln\frac{x}{2a}\right)\right]$$

(8.50)

Similarly, the relative displacements of crack faces behind the tip ($x = \Delta a$) can be obtained from Eq. (8.39) with the results

$$\Delta u_y(\Delta a - x, \pi) = \frac{\sqrt{2(\Delta a - x)}}{4(1 + 4\epsilon^2)\sqrt{\pi}}\left(\frac{\kappa_1 + 1}{\mu_1} + \frac{\kappa_2 + 1}{\mu_2}\right)$$
$$\times [K_1(a + \Delta a)H_1 - K_2(a + \Delta a)H_2]$$

$$\Delta u_x(\Delta a - x, \pi) = \frac{\sqrt{2(\Delta a - x)}}{4(1 + 4\epsilon^2)\sqrt{\pi}}\left(\frac{\kappa_1 + 1}{\mu_1} + \frac{\kappa_2 + 1}{\mu_2}\right)$$
$$\times [K_1(a + \Delta a)H_2 + K_2(a + \Delta a)H_1]$$

(8.51)

where H_1 and H_2 are given in Eq. (8.43) with r replaced by $\Delta a - x$.

Substituting the preceding stresses and displacements into Eq. (8.49), we obtain the energy release rates as follows:

$$G_1 = \frac{1}{32}\left(\frac{\kappa_1 + 1}{\mu_1} + \frac{\kappa_2 + 1}{\mu_2}\right)\left(K_1^2 + K_2^2\right)$$
$$+ \lim_{\Delta a \to 0}\eta\left(1 + \frac{\Delta a}{2a} + \cdots\right)\left[A_R\left(K_1^2 - K_2^2\right) - 2A_I K_1 K_2\right]$$

$$G_2 = \frac{1}{32}\left(\frac{\kappa_1 + 1}{\mu_1} + \frac{\kappa_2 + 1}{\mu_2}\right)\left(K_1^2 + K_2^2\right)$$
$$- \lim_{\Delta a \to 0}\eta\left(1 + \frac{\Delta a}{2a} + \cdots\right)\left[A_R\left(K_1^2 - K_2^2\right) - 2A_I K_1 K_2\right]$$

(8.52)

where

$$\eta = \frac{\cosh(\pi\epsilon)}{8(1 + 4\epsilon^2)\pi}\left(\frac{\kappa_1 + 1}{\mu_1} + \frac{\kappa_2 + 1}{\mu_2}\right)$$

(8.53)

$$A_R = \text{Re}\{A\}$$
$$A_I = -\text{Im}\{A\}$$

$$A = \frac{\sqrt{\pi}}{2} \left(\frac{1}{2} + i\epsilon \right) \left(\frac{\Delta a}{4a} \right)^{-2i\epsilon} \Gamma \left(\frac{1}{2} - i\epsilon \right) / \Gamma(1 - i\epsilon) \qquad (8.54)$$

in which $\Gamma(\ )$ is the Gamma function. It can be seen from these equations that the limits for A, and hence G_1 and G_2 as $\Delta a \to 0$ do not exist for general nonzero oscillation index ϵ. The total energy release rate G defined as $G_1 + G_2$, however, exists and is equal to

$$G = G_1 + G_2 = \frac{1}{16} \left(\frac{\kappa_1 + 1}{\mu_1} + \frac{\kappa_2 + 1}{\mu_2} \right) \left(K_1^2 + K_2^2 \right) \qquad (8.55)$$

which agrees with the result obtained by Malyshev and Salganik [8-5].

8.4.2 Stress Intensity Factor Calculations

While the crack closure technique is convenient and efficient for calculating mixed mode energy release rates and stress intensity factors for homogeneous materials, no convergent energy release rates for Mode I and Mode II exist for interfacial cracks, which makes the conventional crack closure technique futile for stress intensity factor calculations.

To provide a scheme for calculation of stress intensity factors for interfacial cracks based on the energy release rate, Sun and Jih [8-6] and Sun and Qian [8-9] proposed to use modified energy release rates based on a finite crack extension Δa in the finite element calculation of stress intensity factors.

The modified energy release rates have the same expressions as in Eq. (8.49) but without taking the limit of $\Delta a \to 0$, that is,

$$\hat{G}_1 = \frac{1}{2\Delta a} \int_0^{\Delta a} \sigma_{yy}(x,0) \Delta u_y(\Delta a - x, \pi) dx$$

$$\qquad (8.56)$$

$$\hat{G}_2 = \frac{1}{2\Delta a} \int_0^{\Delta a} \sigma_{xy}(x,0) \Delta u_x(\Delta a - x, \pi) dx$$

Substituting Eqs. (8.50) and (8.51) into the preceding equation and neglecting higher order terms in Δa yields

$$\hat{G}_1 = \frac{1}{2} G + C \left[A_R \left(K_1^2 - K_2^2 \right) - 2A_I K_1 K_2 \right]$$

$$\qquad (8.57)$$

$$\hat{G}_2 = \frac{1}{2} G - C \left[A_R \left(K_1^2 - K_2^2 \right) - 2A_I K_1 K_2 \right]$$

where G is the total energy release rate given in Eq. (8.55), A_R and A_I are given in Eq. (8.54), and C is a constant given by

$$C = \frac{\cosh(\pi\epsilon)}{8(1+4\epsilon^2)\pi}\left(\frac{\kappa_1+1}{\mu_1} + \frac{\kappa_2+1}{\mu_2}\right)\left(1+\frac{\Delta a}{2a}\right) \qquad (8.58)$$

By solving the two equations in Eq. (8.57) for K_1 and K_2, and neglecting the negative solution for K_1, Sun and Qian [8-9] obtained the stress intensity factors in terms of $\hat{G}_1$ and $\hat{G}_2$ as follows:

$$K_1 = \sqrt{\frac{4A_R S + 4A_I^2\frac{G}{D} \pm 4\sqrt{\left(A_R S + A_I^2\frac{G}{D}\right)^2 - \left(A_R^2 + A_I^2\right)S^2}}{8\left(A_R^2 + A_I^2\right)}} \qquad (8.59)$$

$$K_2 = \frac{2A_R K_I^2 - S}{2A_I K_I}$$

where

$$S = \frac{\hat{G}_1 - \hat{G}_2}{2C} + \frac{A_R G}{D}$$

$$D = \frac{1}{16}\left(\frac{\kappa_1+1}{\mu_1} + \frac{\kappa_2+1}{\mu_2}\right)$$

Equation (8.59) indicates that, for one set of $\hat{G}_1$ and $\hat{G}_2$, two sets of solutions for K_1 and K_2 may exist. However, there is only one set of K_1 and K_2 that is physically meaningful. The correct K_1 and K_2 can be extracted using conditions on the crack surface displacements.

Use of Eqs. (8.57) and (8.59) to calculate stress intensity factors requires selection of crack extension length Δa. Δa can be selected to satisfy $\Delta a > r_o$, where r_o is the size of the oscillation zone given in Eq. (8.47), so that the normal crack opening is positive. At the same time, the sign of relative tangential crack surface displacement Δu_x can be determined easily from the finite element result. Hence, the displacement conditions for selecting the roots of K_1 and K_2 for interface cracks are

$$\Delta u_y(\Delta a) > 0, \quad sgn(\Delta u_x) \quad \text{determined by FEA}$$

Substitition of Eq. (8.42) into this equation yields the conditions for selecting K_1 and K_2:

$$K_1 H_1 - K_2 H_2 > 0$$

$$sgn(K_1 H_2 + K_2 H_1) = sgn(\Delta u_x) \quad \text{determined by FEA}$$

Note that for the nonoscillation case ($\epsilon = 0$), there is only one pair of K_1 and K_2.

Besides the energy method introduced before, stress intensity factors for interfacial cracks can also be calculated using their relations with the near-tip displacements in Eq. (8.42). Matos et al. [8-13] showed that the stress intensity factors obtained directly from these relations are not reliable. Depending on the location at which these crack surface displacements are taken, the stress intensity factors may vary appreciably. Recognizing that the individual crack surface displacements Δu_x and Δu_y obtained from finite element analysis may not be accurate, but the ratio $\Delta u_x/\Delta u_y$ seems to be more accurate, Sun and Qian [8-9] proposed to use this ratio to determine the stress intensity factor ratio K_2/K_1, which can be expressed in terms of displacement ratio $\Delta u_x/\Delta u_y$ as follows:

$$\frac{K_2}{K_1} = \frac{H_1 - H_2 \times \Delta u_y/\Delta u_x}{H_2 + H_1 \times \Delta u_y/\Delta u_x} \tag{8.60}$$

Stress intensity factors K_1 and K_2 can be obtained from this equation and the total energy release rate Eq. (8.55). Clearly, this displacement ratio scheme is more convenient to perform than the energy method.

Sun and Qian [8-9] calculated the stress intensity factors for a center crack of length $2a$ at the interface between two semi-infinite materials subjected to remote uniform tension σ_{yy}^{∞} and shear σ_{xy}^{∞} as shown in Figure 8.2. In the finite element calculations, the infinite plate was modeled by a panel of size 100×100 with a crack of length 2 ($a = 1$). The finite crack extension was taken as $\Delta a/a = 0.01$. The ratios of Poisson's ratios and Young's moduli of the two materials were $\nu_1/\nu_2 = 1$, and $E_1/E_2 = 10$ and 100. A state of plane stress was assumed.

Table 8.1 shows the finite element results using the energy method and the analytical solutions of Rice and Sih [8-3]. It is found that the average relative errors for K_1 and K_2 are less than 0.1% and 0.6%, respectively. The relatively higher error for K_2 is believed to be purely numerical resulting from a small $\hat{G}_2$ value.

Table 8.2 lists the finite element results using the displacement ratio method and the analytical solutions. Crack surface displacements at various locations were taken

Table 8.1 Stress Intensity Factors for a 100×100 Plane Stress Panel Subjected to Tensile and Shear Loading

		K_1			K_2		
E_1/E_2	$\sigma_{xy}^{\infty}/\sigma_{yy}^{\infty}$	Exact	Energy Method	Error (%)	Exact	Energy Method	Error (%)
	0.0	1.6982	1.6976	0.04	−0.3185	−0.3195	0.30
10	0.5	1.8575	1.8565	0.05	0.5306	0.5319	0.25
	1.0	2.0167	2.0175	0.04	1.3797	1.3711	0.19
	0.0	1.6649	1.6642	0.04	−0.3790	−0.3793	0.08
100	0.5	1.8544	1.8530	0.07	0.4535	0.4564	0.66
	1.0	2.0439	2.0456	0.09	1.2859	1.2815	0.34

Source: Adapted from Sun and Qian [8-9].

Table 8.2 Relative Errors of Stress Intensity Factors Using the Displacement Ratio Method for a 100×100 Plane Stress Panel with $E_1/E_2 = 100$ and $\sigma_{xy}^\infty/\sigma_{yy}^\infty = 1.0$

Element Number	First	Second	Third	Fourth	Fifth
x/a	0.01	0.02	0.03	0.04	0.05
Error for K_1 (%)	2.72	0.67	0.37	0.24	0.13
Error for K_2 (%)	6.46	1.54	0.82	0.48	0.19

Source: Adapted from Sun and Qian [8-9].

to compute the displacement ratios. It is again found that the relative errors for both K_1 and K_2 are less than 1.0% if the nodal displacements were taken at least two elements away from the crack tip.

8.5 FRACTURE CRITERION

In general, fracture of interfaces is inherently of mixed mode due to the material asymmetry. It is known from Chapter 5 that for mixed mode fracture in homogeneous materials, the crack growth path needs to be considered in a fracture criterion. For interface cracks, however, failure often occurs along the interface, that is, the crack growth direction is known. An interface crack may kink into one of the bulk materials if the interface toughness is relatively high, which will be discussed in Section 8.6. Due to the oscillatory nature of the interface crack tip fields, it is more convenient to use energy release rate to formulate fracture criteria for interface failure.

According to the energy release rate criterion, fracture occurs when the energy release rate reaches a critical value. The experiment by Cao and Evans [8-14] shows that the interface toughness for a given bimaterial system is not a constant, but depends on the mode mixity. The energy release rate criterion can thus be expressed as

$$G = G_c(\psi) \tag{8.61}$$

where G_c is the critical energy release rate and ψ is an appropriately defined phase angle. Hutchinson and Suo [8-15] provided some functional forms for the dependence of G_c on the phase angle. For the special case of $\epsilon = 0$ (nonoscillation), ψ may be uniquely defined by the stress intensity factors as follows:

$$\psi = \tan^{-1}\left(\frac{K_2}{K_1}\right) \tag{8.62}$$

For general bimaterial systems ($\epsilon \neq 0$), ψ may not be uniquely defined. This is because, from Eq. (8.31), the ratio of the near-tip shear stress σ_{xy} to the normal stress

σ_{yy} along the interface,

$$\left(\frac{\sigma_{xy}}{\sigma_{yy}}\right)_{\theta=0} = \frac{\mathrm{Im}\left[K_h r^{i\epsilon}\right]}{\mathrm{Re}\left[K_h r^{i\epsilon}\right]}$$

is a function of r, the distance from the crack tip, no matter how small r is. To define the mode mixity unambiguously, Rice [8-10] suggested a definition of stress intensity factor of the classical type denoted by $K_I + iK_{II}$,

$$K_I + iK_{II} = K_h \hat{r}^{i\epsilon} \tag{8.63}$$

where $\hat{r}$ is a reference length. The phase angle according to this stress intensity definition may be expressed as

$$\psi_r = \tan^{-1}\left[\frac{\mathrm{Im}(K_h \hat{r}^{i\epsilon})}{\mathrm{Re}(K_h \hat{r}^{i\epsilon})}\right] \tag{8.64}$$

which represents the mode mixity of the stress field at $r = \hat{r}$ if $\hat{r}$ falls within the K-dominance zone since

$$\psi_r = \tan^{-1}\left(\frac{\sigma_{xy}}{\sigma_{yy}}\right)_{r=\hat{r}}$$

As noted by Rice [8-10] and Hutchinson and Suo [8-15], the reference length $\hat{r}$ may be chosen arbitrarily within a range of length scales. It could be chosen based on the geometrical dimensions of the cracked body or a material length scale. If $\hat{r}$ is chosen as $2a$, the phase angle in Eq. (8.64) becomes

$$\psi_K = \tan^{-1}\left[\frac{\mathrm{Im}(K)}{\mathrm{Re}(K)}\right] = \tan^{-1}\left(\frac{K_2}{K_1}\right) \tag{8.65}$$

where $K = K_1 + iK_2$ is the stress intensity factor defined in Sun and Jih [8-6].

8.6 CRACK KINKING OUT OF THE INTERFACE

In the fracture criterion Eq. (8.61), the crack is assumed to extend along the interface between two dissimilar materials. The crack, however, may also kink into one of the two bulk materials depending on the loading conditions, material properties, and interface fracture toughness. Assume that an interface crack kinks into material 2 as shown in Figure 8.6.

When the kink length Δa is small compared with the original interface crack, the stress intensity factors K_I and K_{II} at the kink crack tip may be expressed in terms of the complex stress intensity factor at the main crack tip [8-16]:

$$K_I + iK_{II} = c(\omega, \alpha, \beta)K_h(\Delta a)^{i\epsilon} + \bar{d}(\omega, \alpha, \beta)\overline{K}_h(\Delta a)^{-i\epsilon} \tag{8.66}$$

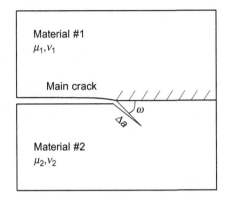

FIGURE 8.6

An interface crack kinking into material 2.

where ω is the kink angle, α and β are two Dundurs' parameters Eq. (8.22), and c and $\overline{d}$ are two dimensionless constants. The energy release rate at the kink crack tip is given by

$$G_{kink} = \frac{\kappa_2 + 1}{8\mu_2} \left(K_I^2 + K_{II}^2 \right)$$

Using the relation Eq. (8.66), G_{kink} can be expressed in the following form:

$$\frac{G_{kink}}{G} = f(\omega, \alpha, \beta, \tilde{\psi}) \tag{8.67}$$

where G is the energy release rate for crack extension along the interface, that is,

$$G = \frac{1}{16} \left(\frac{\kappa_1 + 1}{\mu_1} + \frac{\kappa_2 + 1}{\mu_2} \right) K_h \overline{K}_h$$

and $\tilde{\psi}$ is a phase angle defined by

$$\tilde{\psi} = \psi_r + \epsilon \ln \left(\frac{\Delta a}{\hat{r}} \right)$$

in which ψ_r is the phase angle defined in Eq. (8.64). He and Hutchinson [8-16] gave detailed numerical results of G_{kink}/G.

With the given loading parameter G_{kink}/G, the condition for the interface crack to kink into material 2 at an angle ω may be formulated as

$$\frac{G_{kink}(\omega, \alpha, \beta, \tilde{\psi})}{G} > \frac{\Gamma_c}{G_c(\psi_r)}$$

where Γ_c is the critical energy release rate of material 2 and $G_c(\psi_r)$ is the interface toughness.

8.7 CONTACT AND FRICTION IN INTERFACIAL CRACKS

The oscillatory nature of stress and displacement fields near the tip of an interfacial crack implies contact of the crack surfaces, as discussed in Section 8.3. When the contact zone size is much smaller than the crack length and other in-plane dimensions such as in bimaterial cracks under Mode I loading, the oscillatory crack tip fields may still be used to predict interfacial fracture. However, the contact zone size may become comparable to the crack length if an interfacial crack is subjected to Mode II or shear-dominated loading. In this case, the crack tip fields are significantly altered especially if friction between the crack surfaces is present. This section introduces the fracture mechanics concepts for interfacial cracks considering crack face contact and friction.

8.7.1 Crack Tip Fields

For a crack in homogenous materials with friction or an interface crack without friction, there is always a square root singularity associated with the near-tip field. For an interfacial crack between two dissimilar isotropic materials with friction, however, the stress singularity is always weaker for a stationary crack under monotonic loading, that is, the stress exhibits a singularity of $r^{-\lambda}$ with $\lambda < 0.5$, as originally shown by Comninou [8-7].

The solution with consideration of crack face contact can be obtained basically following the asymptotic expansion procedure introduced by Williams [8-1]. as described in Section 8.1. The Airy stress functions in Comninou's solution are assumed to be in the following form:

$$\phi_j = r^{2-\lambda}F_j(\theta), \quad j = 1,2 \tag{8.68}$$

where (r,θ) are the polar coordinates centered at the crack tip as shown in Figure 8.1. The only difference between the previous form and that of Eq. (8.4) is in the definition of eigenvalue λ. From the stress functions the corresponding stress and displacement fields can be obtained for both materials. Note that with this definition of λ, the near-tip stress field is proportional to $r^{-\lambda}$. The difference in stress singularity between the frictional and frictionless interface cracks is due to the differences in boundary conditions along the crack surfaces.

The boundary conditions along the interface are given as follows. Ahead of the crack tip ($\theta = 0$), the displacements and stresses must be continuous, that is,

$$
\begin{aligned}
(u_r)_1 &= (u_r)_2, & \theta &= 0 \\
(u_\theta)_1 &= (u_\theta)_2, & \theta &= 0 \\
(\sigma_{r\theta})_1 &= (\sigma_{r\theta})_2, & \theta &= 0 \\
(\sigma_{\theta\theta})_1 &= (\sigma_{\theta\theta})_2, & \theta &= 0
\end{aligned}
\tag{8.69}
$$

Behind the crack tip ($\theta = \pm\pi$), the displacements and tractions of the upper and lower crack surfaces must satisfy the following continuity and equilibrium equations:

$$(u_\theta)_1|_{\theta=\pi} = (u_\theta)_2|_{\theta=-\pi}$$
$$(\sigma_{\theta\theta})_1|_{\theta=\pi} = (\sigma_{\theta\theta})_2|_{\theta=-\pi}$$
$$(\sigma_{r\theta})_1|_{\theta=\pi} = (\sigma_{r\theta})_2|_{\theta=-\pi} = -\mu(\sigma_{\theta\theta})_1|_{\theta=\pi} \qquad (8.70)$$
$$(\sigma_{\theta\theta})_1|_{\theta=\pi} \leq 0, \qquad (\sigma_{\theta\theta})_2|_{\theta=-\pi} \leq 0$$

where μ denotes the coefficient of friction. Note that the signs of the shear stresses depend on the direction of slip. To distinguish the directions of slip, the value of μ can also assume negative values.

For instance, a positive μ gives positive shear stresses on both upper and lower surfaces thus indicating the upper medium moves relative to the lower medium in the positive x-direction. A negative μ indicates a slip of the crack surfaces in the opposite direction. These conditions can be expressed in terms of the stress and displacement components with respect to the Cartesian coordinate system as follows:

$$\sigma_{yy}(r,\pm\pi) < 0$$
$$\sigma_{xy}(r,\pm\pi) = -\mu\sigma_{yy}(r,\pm\pi) \qquad (8.71)$$
$$sgn(\mu) = sgn(\Delta u_x)$$

where $\Delta u_x(r) = u_x(r,\pi) - u_x(r,-\pi)$ is the relative displacement in the x-direction between the upper and lower crack surfaces.

The two sets of boundary conditions given by Eqs. (8.69) and (8.70) yield eight homogeneous equations for the eight unknown constants in the solutions for $F_j(\theta)$. For a nontrivial solution to exist, the determinant of the coefficient matrix of the system of eight equations must vanish, that is,

$$\sin^3(\lambda\pi)[\cos(\lambda\pi) - \mu\beta\sin(\lambda\pi)] = 0 \qquad (8.72)$$

where β is a Dundurs' parameter given by Eq. (8.22). Clearly, $\lambda = 0$ is a root of the characteristic Eq. (8.72) and can be ignored as it yields a uniform state of stress. The solution for λ that leads to singular stresses must be obtained from setting the second term in Eq. (8.72) equal to zero with the result

$$\cot(\lambda\pi) = \mu\beta \qquad (8.73)$$

The near-tip stress field along the bimaterial interface and the relative crack surface sliding displacement in plane strain can be calculated from the stress function. In terms of the stress and displacement components with respect to the Cartesian coordinate system we have

$$\sigma_{xy}(r,0) = K_{II}(2\pi r)^{-\lambda}$$
$$\sigma_{xy}(r,\pm\pi) = K_{II}(2\pi r)^{-\lambda}\cos(\lambda\pi)$$
$$\sigma_{yy}(r,0) = 0 \tag{8.74}$$
$$\sigma_{yy}(r,\pm\pi) = -K_{II}\beta(2\pi r)^{-\lambda}\sin(\lambda\pi)$$

and

$$\Delta u_x(r) = u_x(r,\pi) - u_x(r,-\pi) = \frac{\gamma K_{II}\sin(\lambda\pi)}{2(1-\lambda)(2\pi)^\lambda}r^{1-\lambda} \tag{8.75}$$

in which the generalized stress intensity factor is defined as

$$K_{II} = \lim_{r\to 0}(2\pi r)^\lambda \sigma_{xy}(r,0) \tag{8.76}$$

and

$$\gamma = \frac{(3-4\nu_1)(1-\beta)+(1+\beta)}{2\mu_1} + \frac{(3-4\nu_2)(1+\beta)+(1-\beta)}{2\mu_2}$$

The singularity index λ in the stresses Eq. (8.74) is obtained from Eq. (8.73). It is easy to see that λ is 0.5 if β or μ is zero, which corresponds to homogenous materials or the frictionless condition, respectively. If $\mu\beta > 0$, then $\lambda < 0.5$ and the stress singularity is weaker than the inverse square root singularity. On the other hand, $\lambda > 0.5$ if $\mu\beta < 0$, and then a stronger singularity exists.

It was noted by Comninou and Dundurs [8-8] that the problem of an interface crack with friction is a linear process in the context of monotonic loading. In this context, the energy released for a crack extension Δa can be obtained using Irwin's crack closure integral. Because of the contact of crack surfaces, only Mode II fracture mode is present. Sun and Qian [8-17] obtained the total energy release rate associated with a virtual crack extension Δa as follows:

$$\hat{G}(\Delta a) = \frac{1}{2\Delta a}\int_0^{\Delta a}[\sigma_{xy}(r,0) - \sigma_{xy}(\Delta a - r,0)]\Delta u_x(\Delta a - r)dr$$

$$= \frac{\gamma K_{II}^2\sin(\lambda\pi)}{4(1-\lambda)(2\pi)^{2\lambda}}(\Delta a)^{1-2\lambda}\left[\frac{\Gamma(2-\lambda)\Gamma(1-\lambda)}{\Gamma(3-2\lambda)} - \frac{\cos(\lambda\pi)}{2(1-\lambda)}\right] \tag{8.77}$$

in which Γ is the gamma function. The conventional strain energy release rate is defined as

$$G = \lim_{\Delta a\to 0}\hat{G}(\Delta a)$$

In the crack closure integral of Eq. (8.77), the term $\sigma_{xy}(r,0)$ is the interfacial shear stress ahead of the crack tip, and $\sigma_{xy}(\Delta a - r, 0)$ is the frictional shear stress behind the crack tip. During the crack extension of Δa, the shear stress initially ahead of the crack tip reduces to that of the frictional shear stress behind the crack tip after the assumed crack extension. Thus, the energy release rate of Eq. (8.77) can be interpreted as the total energy release rate less the frictional energy dissipation rate.

It is seen from Eq. (8.77) that the energy release rate $\hat{G}(\Delta a)$ vanishes as $\Delta a \to 0$ if $\lambda < 0.5$. Consequently, G cannot be used as a parameter in the fracture criterion. On the other hand, for $\Delta a = \Delta a_0 \neq 0$, $\hat{G}(\Delta a_0)$ is uniquely related to the stress intensity factor K_{II} and, thus, to the near-tip stress field. By selecting a proper characteristic crack closure distance Δa_0, it seems possible to use $\hat{G}(\Delta a_0)$ as a fracture parameter when friction is present.

8.7.2 Finite Element Procedure for Energy Calculation

Analytical solutions for interfacial cracks with frictional sliding are extremely difficult to obtain. Numerical methods such as the finite element method are necessary for practical applications. Sun and Qian [8-17] developed a finite element procedure to calculate the energy release rate $\hat{G}(\Delta a)$ and the energy dissipation due to friction during crack extension. This procedure is based on crack tip nodal forces and displacements. To separate the energy release from the frictional energy dissipation at the crack tip, the elastic nodal force must be separated from the frictional nodal force that results from the frictional traction on the crack surfaces.

For illustration purposes, assume that the crack tip is modeled using four-noded plane strain (or plane stress) elements as shown in Figure 8.7. To facilitate friction as well as nodal release, interface elements INTER2 in the commercial code ABAQUS are used along the interface ahead and behind the crack tip. Let $F_x^{(1)}$ and $F_y^{(1)}$ be the horizontal and vertical nodal forces, respectively, at the crack tip node in the loaded medium just before crack extension. The crack extension is simulated by releasing the crack tip node into two separate nodes a and b, allowing frictional sliding. The relative sliding of the nodes a and b (see Figure 8.7) is denoted by Δu_x.

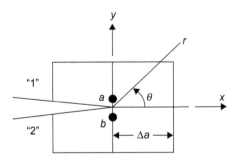

FIGURE 8.7

A crack along a bimaterial interface and the finite elements around the crack tip.

The elastic restoring nodal force at node a before crack extension is $F_x^{(1)} - \mu F_y^{(1)}$. Using Irwin's crack closure concept, the energy release rate for a crack extension of Δa is

$$\hat{G}(\Delta a) = \frac{1}{2\Delta a} \left(F_x^{(1)} - \mu F_y^{(1)} \right) \Delta u_x \tag{8.78}$$

During the assumed crack extension, the total dissipation energy rate associated with crack surface friction is given by

$$G_d(\Delta a) = G_d^N(\Delta a) + G_d^e(\Delta a) \tag{8.79}$$

where $G_d^N(\Delta a)$ is the portion of the dissipation energy rate produced by the newly formed crack surfaces, and $G_d^e(\Delta a)$ is the portion produced by the existing crack surfaces that are in contact. $G_d^N(\Delta a)$ can be determined from the following equation:

$$G_d^N(\Delta a) = \frac{1}{2\Delta a} \left(\mu F_x^{(1)} + \mu F_y^{(2)} \right) \Delta u_x$$

where $F_y^{(2)}$ is the vertical crack tip nodal force after crack extension. The calculation of $G_d^e(\Delta a)$ is similar to that of $G_d^N(\Delta a)$ and should include all the nodes in the contact region before crack extension.

A numerical example presented in Sun and Qian [8-17] is reproduced here. A center crack lying between two dissimilar isotropic media under remote shear loading τ (see Figure 8.8) was used to examine the accuracy of the finite element technique.

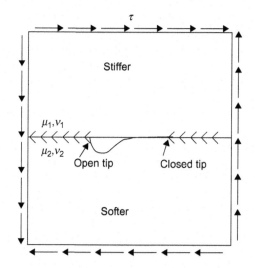

FIGURE 8.8

A cracked bimaterial panel, 20 m × 20 m, under remote shear loading.

Table 8.3 Comparison of Stress Intensity Factor K_{II} from Sun and Qian [8-17] and K_{II}^d Obtained by Comninou and Dunders [8-8] $\left(\overline{K}_{II} = \left(1 + \mu^2\beta^2\right)K_{II}/\tau(2a)^\lambda, \ \mu = 0.5, \ \beta = 0.5 \right)$

$\Delta a/a$	$\overline{K}_{II}$ [8-17]	Error $\left\|\left(K_{II} - K_{II}^d\right)/K_{II}^d\right\|$
3.18×10^{-2}	0.9723	1.28%
1.59×10^{-2}	0.9720	1.25%
7.96×10^{-3}	0.9724	1.29%
3.98×10^{-3}	0.9737	1.42%

The commercial code ABAQUS was used to perform the analysis. The size of the medium was 20 m × 20 m with a crack of 2 m. A similar problem of an infinite bimaterial medium was also investigated by Comninou and Dundurs [8-8] using the elastic dislocation approach. The stress intensity factor obtained by calculating the energy release rate $\hat{G}(\Delta a)$ and using the relation Eq. (8.77) for the closed crack tip (right tip in Figure 8.8) is given together with those from Comninou and Dundurs [8-8] in Table 8.3. The agreement between these two methods is excellent. The slight difference is believed to be due to the finite size effect of the medium considered in Sun and Qian [8-17] in contrast to the infinite medium assumed by Comninou and Dundurs [8-8].

8.7.3 Fracture Criterion

The finite element procedure presented in the previous section is used to study the interfacial crack under shear loading as shown in Figure 8.8. The dimensions of the bimaterial medium are 20 m × 20 m with a half crack length $a = 1$ m. The applied shear stress τ is assumed to be 0.12 MPa. The material constants are $\mu_1 = 35$ MPa, $\mu_2 = 0.35$ MPa, and $v_1 = v_2 = 0$. These material constants yield a Dundurs' parameter of $\beta = 0.49$.

The relative crack surface normal displacements for $\mu = 0.2$ and 0.5, respectively, are shown in Figure 8.9. It is seen that there is a sizable contact zone near the right tip ($x/a = 1$) while the crack is open at the left tip ($x/a = -1$). The length of contact does not seem to be affected much by the coefficient of friction. For $\mu = 0.0, 0.5$, and 1.0 together with $\beta = 0.5$, the contact zone sizes normalized with half crack size are 0.655, 0.76, and 0.83, respectively [8-8]. Theoretically, there is an extremely small contact zone at the open left tip. However, this contact zone is so small that the left crack tip can be treated as an open crack with the classical oscillatory stress field [8-9].

The finite extension strain energy release rate $\hat{G}(\Delta a)$, dissipation energy rate $G_d(\Delta a)$, and total energy release rate $\hat{G}(\Delta a) + G_d(\Delta a)$ for the problem of Figure 8.8 with $\mu = 0.5$ corresponding to various crack extensions are shown in Figures 8.10 and 8.11 for the open (left) and closed (right) crack tips, respectively. At the left crack tip,

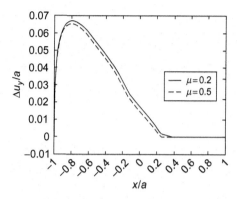

FIGURE 8.9

Relative crack surface normal displacement; a is half crack length (adapted from Sun and Qian [8-17]).

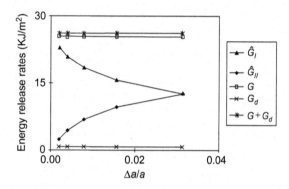

FIGURE 8.10

Finite extension energy release rates at the open crack tip (adapted from Sun and Qian [8-17]).

the contact zone r_c/a is usually around 10^{-5} to 10^{-7} and can be virtually treated as an open crack tip. The modified crack closure technique can therefore be applied to calculate the energy release rates.

It is seen in Figure 8.10 that individual finite extension strain energy release rates $\hat{G}_I$ and $\hat{G}_{II}$ are Δa-dependent, while the total strain energy release rate $G = \hat{G}_I + \hat{G}_{II}$ as well as the dissipation energy rate G_d remain constant for different crack extensions. It is noted that the dissipation energy rate is fairly small compared with energy release rate. This indicates that the frictional effect on the open crack tip is fairly small. It can also be shown that the stress distribution ahead of the open crack tip is very close to those of the oscillatory solution obtained based on the traction-free

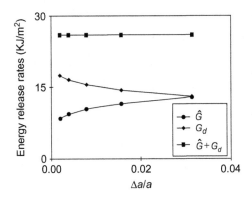

FIGURE 8.11

Finite extension energy release rates at the closed crack tip in the panel shown in Figure 8.8 for $\mu = 0.5$ (adapted from Sun and Qian [8-17]).

crack surface condition. It is concluded that frictional effect at the open crack tip is negligible. All existing approaches for the calculation of fracture parameters for bimaterial interfacial cracks are thus applicable (e.g., see [8-9]).

As for the fracture behavior at the closed crack tip (the right tip), it is clearly seen in Figure 8.11 that the finite extension strain energy release rate $\hat{G}$ (which is Mode II) decreases, and the dissipation energy rate G_d increases, while the total energy release rate $\hat{G} + G_d$ remains constant when crack extension decreases. The numerical results [8-17] indicate that the work done by the external force during the crack extension of Δa is equal to the sum of the increase of the strain energy in the cracked body, the strain energy released $\hat{G}\Delta a$, and the energy dissipated on the crack surfaces $G_d\Delta a$. As $\Delta a \to 0$, the work done is then equal to the gain in the strain energy of the cracked body plus the energy dissipated by friction.

Recall the relationship between energy release rate $\hat{G}$ and crack extension Δa in Eq. (8.77); energy release rate indeed decreases and eventually vanishes when Δa approaches zero due to the weak stress singularity. The energy release rate is therefore Δa-dependent, while the increase of dissipation energy rate makes up the decrease of energy release rate for total energy balance when Δa decreases. The implication of decreasing energy release rate (as $\Delta a \to 0$) is that the classical energy release rate concept is no longer valid because of its vanishing value. Consequently, an energy release rate associated with a finite crack extension is needed to quantify the weak singular stress field ahead of the crack tip.

The decreasing behavior of energy release rate was also noticed by Stringfellow and Freund [8-18] who computed the J-integral for the frictional sliding fracture in a thin film on a substrate. It was pointed out later by Deng [8-19] that the path independence of the J-integral no longer exists due to the crack surface traction resulting from friction, and the J-integral for a vanishingly small contour becomes zero indicating strain energy release rate also vanishes.

In view of the foregoing, Sun and Qian [8-17] proposed a fracture criterion using the energy release rate of a characteristic crack extension as a fracture toughness parameter, that is,

$$\hat{G}(\Delta a_0) = \frac{\Delta W_e}{\Delta a} - G_d(\Delta a_0) - \frac{\Delta U}{\Delta a} = \hat{G}_c \qquad (8.80)$$

where ΔW_e, ΔU, and $G_d(\Delta a_0)$ are the external work done, strain energy change, and the dissipation energy rate for a crack extension of Δa_0, respectively. It is seen from Eq. (8.77) that for a fixed Δa_0 the finite extension energy release rate $\hat{G}$ has a unique relation with the generalized stress intensity factor K_{II}. The fracture criterion given by Eq. (8.80) states that the interfacial crack would grow if the near-tip stress field reaches a certain critical state.

8.7.4 Effect of Compressive Loading

An open crack may be forced to close if compressive loads normal to the crack surfaces are applied in addition to the shear load. An example is the residual stresses between the fiber and matrix in a composite resulting from the mismatch of coefficients of thermal expansion. Theoretically, the open region of the crack cannot be completely closed by externally applied compression. However, if the open crack is almost closed due to compressive loads, then the associated near-tip stress field is dominated by a singularity stronger than $\lambda = 0.5$, as discussed by Qian and Sun [8-20].

For this type of "closed" interfacial crack, the strain energy release rate for a finite extension Δa is also given by Eq. (8.77). It is obvious that if $\lambda > 0.5$, then $\hat{G}$ becomes unbounded as $\Delta a \to 0$. Consequently, the intrinsic interfacial fracture toughness must be measured in terms of the finite extension strain energy release rate $\hat{G}$ for a characteristic distance a_0. However, the toughnesses of the two closed interfacial cracks with stronger and weaker singularities, respectively, should not be interchanged, because they correspond to different critical near-tip stress states.

References

[8-1] M.L. Williams, The stresses around a fault or crack in dissimilar media, Bull. Seismol. Soc. Am. 49 (1959) 199–204.

[8-2] J. Dundurs, Edge-bonded dissimilar orthogonal elastic wedges under normal and shear loading, J. Appl. Mech. 36 (1969) 650–652.

[8-3] J.R. Rice, G.C. Sih, Plane problems of cracks in dissimilar media, J. Appl. Mech. 32 (1965) 418–423.

[8-4] J.W. Hutchinson, M. Mear, J.R. Rice, Crack paralleling an interface between dissimilar materials, J. Appl. Mech. 54 (1987) 828–832.

[8-5] B.M. Malysev, R.L. Salganik, The strength of adhesive joints using the theory of cracks, Int. J. Fract. 1 (1965) 114–127.

[8-6] C.T. Sun, C.J. Jih, On strain energy release rate for interfacial cracks in bimaterial media, Eng. Fract. Mech. 28 (1987) 13–20.

[8-7] M. Comninou, Interface crack with friction in the contact zone, J. Appl. Mech. 44 (1977) 780–781.

[8-8] M. Comninou, J. Dundurs, Effect of friction on the interface crack loaded in shear, J. Elast. 10 (1980) 203–212.

[8-9] C.T. Sun, W. Qian, The use of finite extension strain energy release rates in fracture of interfacial cracks, Int. J. Sol. Struct. 34 (1997) 2595–2609.

[8-10] J.R. Rice, Elastic fracture mechanics concepts for interficial cracks, J. Appl. Mech. 55 (1988) 98–103.

[8-11] A.K. Gautsen, J. Dundurs, The interface crack in a tension field, J. Appl. Mech. 54 (1987) 93–98.

[8-12] G.R. Irwin, Analysis of stresses and strains near the end of a crack traversing a plate, J. Appl. Mech. 24 (1957) 361–364.

[8-13] P.P.L. Matos, R.M. McMeeking, P.G. Charalamibides, D. Drory, A method for calculating stress intensities in bimaterial fracture, Int. J. Fract. 40 (1989) 235–254.

[8-14] H.C. Cao, A.G. Evans, An experimental study of the fracture—resistance of bimaterial interfaces, Mech. Mater. 7 (1989) 295–304.

[8-15] J.W. Hutchinson, and Z. Suo, Mixed mode cracking in layered materials, Adv. Appl. Mech. 29 (1992) 63–187.

[8-16] M.-Y. He, J.W. Hutchinson, Kinking of a crack out of an interface, J. Appl. Mech. 56 (1989) 270–278.

[8-17] C.T. Sun, W. Qian, A treatment of interfacial cracks in the presence of friction, Int. J. Fract. 94 (1998) 371–382.

[8-18] R.G. Stringfellow, L.B. Freund, The effect of interfacial friction on the buckle-driven spontaneous delamination of a compressed thin film, Int. J. Sol. Struct. 30 (1993) 1379–1395.

[8-19] X. Deng, Mechanics of debonding and delamination in composites: Asymptotic studies, Compos. Eng. 5 (1995) 1299–1315.

[8-20] W. Qian, C.T. Sun, Frictional interfacial crack under combined shear and compressive loading, Compos. Scie. Technol. 58 (1998) 1753–1761.

PROBLEMS

8.1 Calculate the two Dundurs' parameters for the following two pairs of materials: (1) Glass/epoxy with $E_1 = 70$ GPa, $\nu_1 = 0.2$ (glass) and $E_2 = 2$ GPa, $\nu_2 = 0.4$ (epoxy); (2) Al$_2$O$_3$/aluminum with $E_1 = 350$ GPa, $\nu_1 = 0.25$ (Al$_2$O$_3$) and $E_2 = 70$ GPa, $\nu_2 = 0.33$ (aluminum).

8.2 Derive the stress intensity factors for a semi-infinite interface crack subjected to uniform pressure p along part of the crack faces as shown in Figure 8.12.

8.3 Derive the relationship between the energy release rate and the stress intensity factor for Mode III interface cracks.

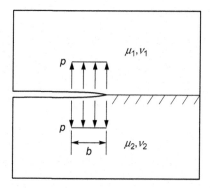

FIGURE 8.12

A semi-infinite interface crack subjected to pressure p along part of the crack faces.

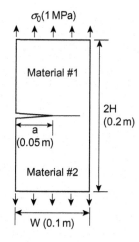

FIGURE 8.13

An edge interface crack in a bimaterial plate subjected to uniform tension.

8.4 Use plane strain or plane stress two-dimensional finite element to model the plate with an interfacial crack loaded as shown in Figure 8.13 The material properties of Material #1 and Material #2 are $E_1 = 210$ GPa, $\nu_1 = 0.3$ and $E_2 = 210$ GPa, $\nu_2 = 0.1$, respectively. Calculate G_I and G_{II} using the modified crack closure method. Check the convergence and the crack surface displacement near the crack tip.

Cohesive Zone Model

To a beginner, the most difficult concept to accept in the linear elastic fracture mechanics (LEFM) is the fact that stresses at the crack tip are infinitely large, or singular. This stress singularity results from the linear elastic continuum representation of materials and the assumption of a perfectly sharp crack. Since stress singularity contradicts our intuition about failure of materials, there have been efforts in introducing ways to remove the stress singularity at the crack tip. Adopting inelastic behavior in solids is a natural extension of the LEFM to avoid stress singularity. Irwin's plastic zone adjustment approach as discussed in Chapter 6 belongs to this category. However, such an approach often goes back to using the parameters employed by the LEFM such as stress intensity factor.

A somewhat different approach toward the same objective is the cohesive zone model, which adds a zone of vanishing thickness ahead of the crack tip with the intention of describing more realistically the fracture process without the use of stress singularity. The cohesive zone is idealized as two cohesive surfaces, which are held together by a cohesive traction. The material failure is characterized by the complete separation of the cohesive surfaces and the separation process is described by a cohesive law that relates the cohesive traction and the relative displacement of the cohesive surfaces. Hence, a physical crack extension occurs when the separation displacement at the tail of the cohesive zone (physical crack tip) reaches a critical value. One of the key advantages offered by the cohesive zone model is that it has an intrinsic fracture energy dissipation mechanism in contrast to the classical continuum based fracture mechanics for which such a mechanism is absent.

9.1 THE BARENBLATT MODEL

Barenblatt [9-1] proposed a cohesive zone concept to study the fracture of brittle materials, aiming to introduce a separation mechanism at the atomic level to describe the actual separation of materials, and to eliminate the crack tip stress singularity. Barenblatt believed that the crack surfaces at the tip region are so close that they can be treated as two atomic layers held by the atomic bonding forces, as shown in

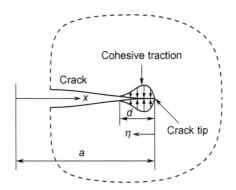

FIGURE 9.1

A Barenblatt cohesive zone near a crack tip.

Figure 9.1. The Barenblatt theory of brittle fracture is based on the following three basic hypotheses:

1. There exists a cohesive zone near the crack tip where the upper and lower crack faces are held by the so-called cohesive traction, which has a magnitude on the order of up to the theoretical strength of the solid. The size of the cohesive zone is much smaller than the crack length.

2. The size of the cohesive zone, and the distribution and magnitude of the cohesive traction at the onset of crack growth are independent of crack geometries and external loads.

3. Stresses are finite everywhere including the crack tip, that is, no stress singularity exists at the crack tip. The first two assumptions were explicitly described in the original paper of Barenblatt [9-1]. The third assumption was actually used in the derivation. Note that the cohesive zone in the Barenblatt model is actually part of the crack.

Now we use a Mode I crack in an infinite elastic solid as an example to illustrate the Barenblatt theory for brittle solids. Consider a through crack of length $2a$ in a two-dimensional infinite elastic solid subjected to uniform tension, σ_∞, at infinity. According to the first assumption of the Barenblatt theory, a small part of the crack faces, called the cohesive zone, near each tip is acted on by the cohesive traction, as shown in Figure 9.1. The size of the cohesive zone, d, is much smaller than the crack length, that is, $d/a << 1$. The distribution of the cohesive traction is represented by

$$\sigma = g(x), \quad a - d \leq |x| \leq a$$

where σ is the cohesive traction.

In the Barenblatt model, the bulk material is still regarded as an elastic continuum. Consequently, the superposition method can be used to treat the problem. According to LEFM, the normal stress at the crack line near the crack tip due to the remote tension is given by (see Eqs. (3.41) and (3.44) in Chapter 3)

$$\sigma_{yy}^{(1)} = \frac{K_I}{\sqrt{2\pi(x-a)}}, \quad x > a \tag{9.1}$$

where $K_I = \sigma_\infty \sqrt{\pi a}$ is the stress intensity factor. Meanwhile, the normal stress ahead of the crack tip induced by the cohesive traction $g(x)$ can be written as

$$\sigma_{yy}^{(2)} = \frac{-B}{\sqrt{2\pi(x-a)}}, \quad x > a \tag{9.2}$$

where B may be obtained using the stress intensity factor formula for concentrated forces and the superposition method described in Eq. (3.79) in Chapter 3:

$$B = 2\sqrt{\frac{a}{\pi}} \int_{a-d}^{a} \frac{g(\xi)d\xi}{\sqrt{a^2 - \xi^2}}$$

Here B is actually $-K_I$ in Eq. (3.79) and $g(\xi)$ is symmetric. By using the first assumption $d/a << 1$, this equation becomes

$$B = \sqrt{\frac{2}{\pi}} \int_{0}^{d} \frac{g(\eta)d\eta}{\sqrt{\eta}} \tag{9.3}$$

where $\eta = a - \xi$. The total normal stress near the crack tip thus becomes

$$\sigma_{yy} = \sigma_{yy}^{(1)} + \sigma_{yy}^{(2)} = \frac{K_I - B}{\sqrt{2\pi(x-a)}}, \quad x > a \tag{9.4}$$

The third assumption in the Barenblatt theory now requires the removal of the stress singularity at the crack tip. This is fulfilled by requiring that

$$K_I - B = 0 \tag{9.5}$$

that is, B always equals K_I.

Up to this point, we have said nothing about the magnitude and the distribution of the cohesive traction, and the size of the cohesive zone except that it is small. Because K_I increases with the applied load, the size of the cohesive zone and the magnitude of the cohesive traction will also change with the load to maintain the nonsingularity of stresses at the crack tip (Eq. (9.5)).

According to the second assumption, the size of the cohesive zone and the distribution of the cohesive traction at the onset of crack growth are independent of applied load and crack size. That is, d and $g(\eta)$ in Eq. (9.3) at crack initiation are material properties. Thus, the parameter B should remain constant during crack extension. This constant value can be regarded as a material constant and can be rewritten as

$$B = B_c = \sqrt{\frac{2}{\pi}} \int_0^{d_c} \frac{g_c(\eta)d\eta}{\sqrt{\eta}} \qquad (9.6)$$

where $g_c(\eta)$ and d_c are the cohesive traction distribution and the cohesive zone size at the onset of crack growth. In view of Eq. (9.5), the fracture criterion in terms of B_c according to the Barenblatt model is equivalent to Irwin's LEFM criterion in terms of K_{Ic}.

In the Barenblatt model, the specific form of the cohesive traction distribution is not given. Rather, the cohesive traction is related to the separation of atomic layers. A fundamental question arises as to how the cohesive traction on the order of the theoretical strength can be incorporated into an elastic continuum. The Barenblatt model avoids the stress singularity at the crack tip, but creates another problem in linking different scales (atomic to continuum).

It is noted that in the Barenblatt model, the cohesive zone represents a small part of the crack surfaces behind the crack tip, which is in contrast with the currently prevailing concept that the cohesive zone actually represents the fracture process ahead of the crack.

9.2 COHESIVE ZONE CONCEPT IN CONTINUUM MECHANICS AND COHESIVE LAWS

Since the Barenblatt model was introduced, use of the cohesive zone model to treat fracture problems has gained much attention. Besides providing a more realistic feature in fracture mechanism, this model also renders simplicity in simulations of complex cracking processes. From the continuum mechanics point of view, crack extension embodies complex failure processes at the microscopic level, for example, void nucleation, growth and coalescence in ductile metals, microcracking in ceramics, crazing in certain polymers, and so on.

If the failure process is confined in a narrow band, such as crazing in polymers (see Figure 9.2(a)) and necking in ductile thin-sheet materials (see Figure 9.2(b)), the cohesive zone may be used to represent the narrow deformation band. If such a distinct narrow deformation band is not present, the cohesive zone can only be regarded as a hypothesis or an approximate representation of the crack tip failure process zone, such as microcracking in brittle materials (see Figure 9.3(a)) and void growth and coalescence in ductile metals (see Figure 9.3(b)).

In general, it is assumed in the cohesive zone approach that, ahead of the physical crack tip, there exists a cohesive zone, which consists of upper and lower surfaces called cohesive surfaces held by the cohesive traction. The cohesive traction is related

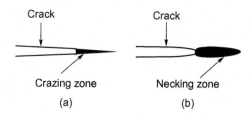

Crack   Crack

Crazing zone   Necking zone

(a)   (b)

FIGURE 9.2

(a) Crazing zone ahead of a crack in a polymer, (b) necking zone in a ductile thin-sheet material.

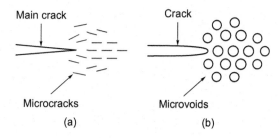

Main crack   Crack

Microcracks   Microvoids

(a)   (b)

FIGURE 9.3

(a) Microcracking zone ahead of a crack in a brittle solid, (b) voids in a ductile metal.

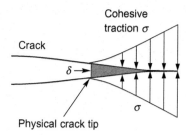

Cohesive
traction σ

Crack

$\delta \rightarrow$

σ

Physical crack tip

FIGURE 9.4

A cohesive zone ahead of a crack tip.

to the separation displacement between the cohesive surfaces by a "cohesive law." Upon the application of external loads to the cracked body, the two cohesive surfaces separate gradually, leading to physical crack growth when the separation of these surfaces at the tail of the cohesive zone (physical crack tip) reaches a critical value. Figure 9.4 depicts a cohesive zone ahead of a crack where σ is the cohesive traction and δ is the separation displacement of the cohesive surfaces.

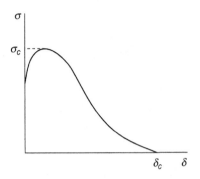

FIGURE 9.5

A general cohesive law describing the relationship between the cohesive traction and the separation displacement.

A cohesive law describes the relationship between the cohesive traction and the separation displacement as follows:

$$\sigma = \sigma_c f(\delta/\delta_c) \tag{9.7}$$

where σ_c is the peak cohesive traction, δ_c a characteristic separation, and f a dimensionless function describing the "shape" of the cohesive traction-separation curve (cohesive curve). The function f depends on the failure mechanism operative either at the microscopic or macroscopic level. Figure 9.5 shows the schematic of the cohesive law Eq. (9.7).

Besides the cohesive parameters σ_c and δ_c, we often use the cohesive energy density Γ_c, or the work of separation per unit area of cohesive surface, defined by

$$\Gamma_c = \int_0^{\delta_c} \sigma(\delta) d\delta \tag{9.8}$$

The cohesive energy density is just the area under the cohesive curve. Once the shape of the cohesive curve is given, only two of the parameters σ_c, δ_c, and Γ_c are independent. When a cohesive zone is based on a distinct narrow deformation band, the cohesive law may be developed by considering the stress and deformation states in the narrow band. For example, Jin and Sun [9-2] derived a cohesive law based on crack front necking in ductile thin-sheet materials. Otherwise, a phenomenological cohesive law is often used. In this case, the cohesive zone parameters such as peak cohesive traction and cohesive energy density need to be calibrated by experiment. Some commonly used phenomenological cohesive laws are given next.

9.2.1 **The Dugdale Model**

The Dugdale model discussed in Chapter 6 for ductile metals may be regarded as a cohesive zone model. The strip plastic zone is now the cohesive zone and the yield stress is the cohesive traction. The cohesive law is thus given by (see Figure 9.6(a))

$$\sigma = \sigma_c \tag{9.9}$$

Since the cohesive traction is a constant, we need to specify a critical separation displacement, or a cohesive energy density. The relation between them is

$$\Gamma_c = \sigma_c \delta_c \tag{9.10}$$

In using the Dugdale model to simulate fracture propagation, complete separation of materials occurs when $\delta = \delta_c$.

9.2.2 **A Linear Softening Model**

A linear softening cohesive zone model is often used to describe the progressive failure in some quasibrittle materials, for example, ceramics and concrete. This cohesive law is expressed in the form (see Figure 9.6(b))

$$\sigma = \sigma_c \left(1 - \delta/\delta_c\right) \tag{9.11}$$

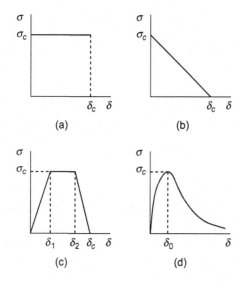

FIGURE 9.6

Cohesive zone models: (a) the Dugdale model, (b) a linear softening model, (c) a trapezoidal model, and (d) an exponential model.

The cohesive energy density is

$$\Gamma_c = \frac{1}{2}\sigma_c\delta_c \tag{9.12}$$

The cohesive law Eq. (9.11) shows that the cohesive traction attains an initial peak value σ_c, decreases with increasing separation displacement, and vanishes at the critical separation of δ_c.

9.2.3 A Trapezoidal Model

The trapezoidal model is an extension of the Dugdale model Eq. (9.9) and the linear softening model Eq. (9.11) with the following traction-separation relationship (Figure 9.6(c)):

$$\sigma = \begin{cases} \sigma_c\,(\delta/\delta_1), & 0 \le \delta \le \delta_1 \\ \sigma_c, & \delta_1 \le \delta \le \delta_2 \\ \sigma_c(\delta_c - \delta)/(\delta_c - \delta_2), & \delta_2 \le \delta \le \delta_c \end{cases} \tag{9.13}$$

where δ_1 and δ_2 are two parameters. The cohesive law Eq. (9.13) can represent a Dugdale model ($\delta_1 = 0$ and $\delta_2 = \delta_c$), a linear softening model ($\delta_1 = \delta_2 = 0$), and their combinations by appropriately selecting the parameters δ_1 and δ_2. The cohesive energy density Γ_c is now given by

$$\Gamma_c = \frac{1}{2}\sigma_c(\delta_c + \delta_2 - \delta_1) \tag{9.14}$$

The cohesive law given in Eq. (9.13), however, is not directly derived from a failure mechanism.

9.2.4 An Exponential Model

An exponential cohesive zone model based on the universal binding energy curve from the atomistic consideration (Rose et al. [9-3]) has found frequent usage in simulating crack growth in ductile metals. The model takes a computationally convenient exponential form (see Figure 9.6(d)),

$$\sigma = \sigma_c\left(\frac{\delta}{\delta_0}\right)\exp\left(1 - \frac{\delta}{\delta_0}\right) \tag{9.15}$$

where δ_0 is the separation displacement at which $\sigma = \sigma_c$, the peak cohesive traction. The cohesive energy density of model Eq. (9.15) is

$$\Gamma_c = \int_0^{\infty} \sigma\,d\delta = e\sigma_c\delta_0 \tag{9.16}$$

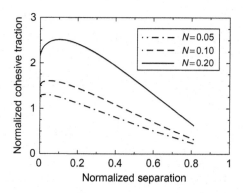

FIGURE 9.7

A cohesive law based on necking (adapted from Jin and Sun [9-2]).

where $e = \exp(1)$. Theoretically, this cohesive energy can be completely dissipated only when the separation displacement becomes unbounded. Practically, the complete separation is assumed to occur when the separation displacement becomes sufficiently larger than δ_0, say, $6\delta_0$. At this point, the cohesive energy dissipated is greater than 95% of Γ_c, and the cohesive traction is less than 5% of the peak value σ_c.

9.2.5 A Cohesive Zone Model Based on Necking

Based on the crack front necking phenomenon in ductile thin-sheet materials [9-4], Jin and Sun [9-2] derived an implicit relationship between the cohesive traction and separation displacement. Figure 9.7 shows the numerical results of the cohesive law for different values of the (power law) hardening exponent N. The cohesive traction is normalized by the yield stress and the separation is normalized by the sheet thickness. It can be seen from the figure that the cohesive traction first increases from a nonzero finite value at the beginning of separation, quickly reaches the peak, and then decreases with further increasing separation displacement. The peak cohesive traction varies from 1.15 times the yield stress for perfectly plastic materials to about 2.5 times the yield stress for modest hardening materials (hardening exponent of 0.2).

9.3 A DISCUSSION ON THE LINEAR HARDENING LAW

The linear hardening traction—separation relationship is a basic element in several widely used cohesive laws. For example, the trapezoidal model Eq. (9.13) reduces to the linear hardening model when the opening displacement of the cohesive surfaces has not reached δ_1. In numerical simulations, a linear hardening model is usually used as the first step in the nonlinear exponential model Eq. (9.15). The linear hardening model itself has also been used in some studies. These kind of models have a common feature that the cohesive traction takes a zero value at the start of cohesive surface separation (zero separation).

In other words, these models have an initial (or asymptotically initial) linear elastic response. Physically speaking, a cohesive zone model describes the fracture process in a material and the material separation occurs only after it deforms significantly. It is thus expected that cohesive models should have a finite traction at the start of separation, or an initial rigid response. Jin and Sun [9-2] studied the problem from the viewpoint of stress singularity cancellation at the cohesive zone tip.

Consider a two-dimensional elastic medium of infinite extent with a crack of length $2a$ subjected to remote tension p, as shown in Figure 9.8. A cohesive zone of length $\rho = c - a$ develops ahead of each crack tip upon external loading. It is assumed that the cohesive traction σ and the separation δ follow the linear hardening model,

$$\sigma = \begin{cases} \sigma_c \, (\delta/\delta_c), & 0 \le \delta \le \delta_c \\ 0, & \delta > \delta_c \end{cases} \tag{9.17}$$

The cohesive zone modeling approach employs the cohesive crack assumption that the cohesive zone is treated as an extended part of the crack, with the total stress intensity factor vanishing at the tip of the cohesive zone. The stress intensity factor due to the applied load p is

$$K_I^{app} = p\sqrt{\pi c}$$

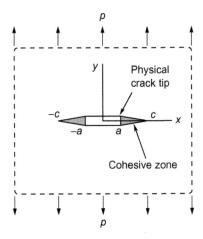

FIGURE 9.8

Cohesive zones ahead of a crack in an infinite elastic solid.

and the stress intensity factor due to the cohesive traction σ may be obtained using the superposition method and the basic solution described in Chapter 3 as follows:

$$K_I^{coh} = -\frac{2}{\sqrt{\pi c}} \int_a^c \frac{\sigma(\xi)d\xi}{\sqrt{1 - \xi^2/c^2}}$$

To remove the stress singularity at the cohesive zone tip $(x = c)$, we have

$$K_I^{app} + K_I^{coh} = p\sqrt{\pi c} - \frac{2}{\sqrt{\pi c}} \int_a^c \frac{\sigma(\xi)d\xi}{\sqrt{1 - \xi^2/c^2}} = 0 \tag{9.18}$$

At the same time, the crack/cohesive surface opening displacement δ is

$$\delta(x) = \frac{4cp}{E^*}\sqrt{1 - \frac{x^2}{c^2}} - \frac{4}{\pi E^*} \int_a^c G(x, \xi)\sigma(\xi)d\xi \tag{9.19}$$

where the first and second terms on the right side are due to the external load p and the cohesive traction σ, respectively, $E^* = E$ for plane stress and $E^* = E/(1 - v^2)$ for plane strain, and $G(x, \xi)$ is a function given by

$$G(x, \xi) = \ln\left|\frac{\sqrt{1 - x^2/c^2} + \sqrt{1 - \xi^2/c^2}}{\sqrt{1 - x^2/c^2} - \sqrt{1 - \xi^2/c^2}}\right| \tag{9.20}$$

Substitution of model Eq. (9.17) into Eq. (9.19) and use of Eq. (9.18) lead to the integral equation for the opening displacement δ:

$$\delta(x) + \frac{4\sigma_c}{\pi E^*\delta_c} \int_a^c K(x, \xi)\delta(\xi)d\xi = 0, \quad a \le x \le c \tag{9.21}$$

where $K(x, \xi)$ is the kernel given by

$$K(x, \xi) = G(x, \xi) - 2\sqrt{\frac{1 - x^2/c^2}{1 - \xi^2/c^2}}, \quad a \le x, \xi \le c$$

Equation (9.21) is a homogeneous linear integral equation for δ and has only a trivial solution. To prove this, we add a nonhomogeneous term,

$$\frac{4cK_{Ic}}{E^*\sqrt{\pi c}}\sqrt{1 - \left(\frac{x}{c}\right)^2}$$

to the right side of Eq. (9.21) (where K_{Ic} is a positive constant) to obtain the following equation:

$$\delta(x) + \frac{4\sigma_c}{\pi E^* \delta_c} \int_a^c K(x,\xi)\delta(\xi)d\xi = \frac{4cK_{Ic}}{E^*\sqrt{\pi c}}\sqrt{1 - \left(\frac{x}{c}\right)^2}, \quad a \leq x \leq c$$

This equation is the integral equation for a crack partially bridged by linear springs. For a general set of cohesive and elasticity parameters $(\sigma_c, \delta_c, E^*)$, the problem has a unique nontrivial solution according to the elasticity theory (Rose [9-5]). This, in turn, requires that its homogeneous equation (9.21) have only a trivial solution [9-6]. The nonexistence of a nontrivial solution for Eq. (9.21) implies that the crack tip singularity can not be canceled. As a result, the cohesive zone model can not assume a linear hardening law with an initial zero cohesive traction, if the stress singularity is to be removed, or if the energy dissipation is not allowed, at the tip of the cohesive zone.

Although the preceding discussion is limited to an infinite plate with a central crack, the conclusion should apply to the general geometric cases and should also apply to the trapezoidal model Eq. (9.13) as well as nonlinear cohesive zone models with $\sigma = 0$ at $\delta = 0$ because the initial nonlinear hardening can be approximated by a linear hardening relationship near $(\sigma, \delta) = (0,0)$. These kind of nonlinear models include the widely used model Eq. (9.15).

Because the stress singularity can not be removed with a cohesive zone model having an initial linear elastic response, the cohesive zone will not be able to develop naturally, but has to be given a priori. Jin and Sun [9-7] examined the energy dissipation at the tip of a prescribed cohesive zone using a bilinear model ($\delta_1 = \delta_2$ in Figure 9.6(c)). They concluded that the energy dissipation at the cohesive zone tip may be neglected only when the initial stiffness of the cohesive model is appropriately high and the pre-embedded cohesive zone is set much greater than the cohesive zone size with canceled stress singularity at the cohesive zone tip.

9.4 COHESIVE ZONE MODELING AND LEFM

To a certain extent, it is true that the cohesive zone fracture model was introduced partially to remove the stress singularity in the classical continuum crack model and partially to incorporate the physically more realistic material separation mechanisms at the atomic scale. The relation between these two different fracture models is certainly a topic of interest.

We discuss the relationship using the J-integral approach proposed by Rice [9-8]. Consider a cohesive zone ahead of the crack tip as shown in Figure 9.9. Assume that the cohesive zone is so small that the K-field near the crack tip is not disturbed. Evaluating the J-integral along a contour Γ_1 completely embedded in the K-dominance

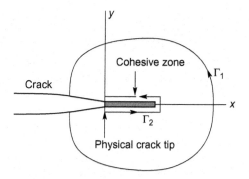

FIGURE 9.9

Integration contours around a crack tip with a cohesive zone.

zone at the (physical) crack initiation state, we have

$$J^{(1)} = \frac{K_{Ic}^2}{E^*} = G_{Ic} \tag{9.22}$$

where K_{Ic} is the fracture toughness and G_{Ic} is the critical energy release rate in LEFM. Alternatively, we may evaluate the J-integral along a contour Γ_2 that consists of the upper and lower surfaces of the cohesive zone. Since $dy = 0$ along the integration path, the J-integral becomes, at the crack initiation state,

$$J^{(2)} = -\int_{\Gamma_2} \sigma_{ij} n_j \frac{\partial u_i}{\partial x_1} dl = -\int_0^{L_c} \sigma(\delta) \frac{d\delta}{dx} dx$$

where L_c is the cohesive zone length at crack initiation. Noting that $\delta = 0$ at the tip of the cohesive zone and $\delta = \delta_c$ at the physical crack tip at crack initiation, the equation can be reduced to

$$J^{(2)} = \int_0^{\delta_c} \sigma(\delta) d\delta = \Gamma_c \tag{9.23}$$

Because no field singularities or discontinuities exist between the contours Γ_1 and Γ_2, the J-integrals evaluated along the two contours should be equal, that is,

$$\Gamma_c = \frac{K_{Ic}^2}{E^*} = G_{Ic} \tag{9.24}$$

This relationship indicates that the cohesive zone model and the LEFM are equivalent when the cohesive energy density is identified as the critical energy release rate in the LEFM and the cohesive zone is embedded in the K-dominance zone.

9.5 COHESIVE ZONE MODELING OF INTERFACIAL FRACTURE

From a physical point of view, the cohesive zone approach appears particularly appropriate for investigating interface fracture because the interface region may be modeled as a cohesive zone. There are, however, a few conceptual issues that need to be examined carefully. For example, the cohesive zone length for Mode I fracture can be determined from the condition that no stress singularities exist at the cohesive zone tip. For interface fracture, however, there are generally two independent cohesive tractions (normal and shear) and a single cohesive zone length may not be able to satisfy the condition that stress singularity at the cohesive zone tip be canceled. It is known from Chapter 8 that interfacial fracture toughness is not a constant but is a function of mode mixity. Such a characteristic of interfacial cracks must also be accounted for by the cohesive zone model. These issues are discussed in this section.

9.5.1 Mixed Mode Cohesive Law

It is known from Chapter 8 that the opening and sliding modes are inherently coupled together in fracture along bimaterial interfaces due to material property asymmetry. When using cohesive zone models to study interface fracture, mixed mode cohesive traction—separation relations are thus required. For mixed opening/sliding mode fracture, there are two separation components across the cohesive surfaces: the opening separation δ_n and the sliding component δ_s. An extension of Eq. (9.7) to the mixed mode cohesive zone model takes the following general form:

$$\sigma_n = f_n(\delta_n, \delta_s)$$
$$\sigma_s = f_s(\delta_n, \delta_s) \tag{9.25}$$

where σ_n and σ_s are the normal and shear cohesive tractions, respectively.

The functional forms of f_n and f_s in Eq. (9.25), or the shapes of the cohesive curves, have not been well understood largely due to the opening/sliding coupling effects. A cohesive energy potential is often used to simplify the procedure of determining f_n and f_s. For example, Needleman [9-9] assumed the existence of a cohesive energy potential $\Phi(\delta_n, \delta_s)$ as a function of two separations δ_n and δ_s and derived the mixed mode cohesive law as follows:

$$\sigma_n = \frac{\partial \Phi(\delta_n, \delta_s)}{\partial \delta_n}$$
$$\sigma_s = \frac{\partial \Phi(\delta_n, \delta_s)}{\partial \delta_s} \tag{9.26}$$

Ortiz and Pandolfi [9-10] defined an effective cohesive traction σ_{eff} and an effective separation δ_{eff}

$$\sigma_{eff} = \sqrt{\sigma_n^2 + \eta^{-2}\sigma_s^2}$$
$$\delta_{eff} = \sqrt{\delta_n^2 + \eta^2\delta_s^2} \tag{9.27}$$

where η is a coefficient, and assumed that the cohesive energy potential $\Phi(\delta_{eff})$ is a function of the effective separation δ_{eff}. The effective cohesive traction is then derived from the potential $\Phi(\delta_{eff})$ by

$$\sigma_{eff} = \frac{d\Phi(\delta_{eff})}{d\delta_{eff}} \tag{9.28}$$

and the cohesive tractions can be obtained as

$$\sigma_n = \frac{\partial \Phi}{\partial \delta_n} = \frac{d\Phi}{d\delta_{eff}} \frac{\partial \delta_{eff}}{\partial \delta_n} = \frac{\sigma_{eff}}{\delta_{eff}} \delta_n$$

$$\sigma_s = \frac{\partial \Phi}{\partial \delta_s} = \frac{d\Phi}{d\delta_{eff}} \frac{\partial \delta_{eff}}{\partial \delta_s} = \eta^2 \frac{\sigma_{eff}}{\delta_{eff}} \delta_s \tag{9.29}$$

It follows from this equation that $\sigma_n/\sigma_s = \eta^{-2}\delta_n/\delta_s$ which implies a kind of proportional deformation. Tvergaard and Hutchinson [9-11] used a different form of the effective separation and their cohesive traction–separation relations are similar to those in Eq. (9.29).

In general, the cohesive energy density can be calculated from

$$\Gamma_c = \int_0^{\delta_n^c} \sigma_n d\delta_n + \int_0^{\delta_s^c} \sigma_s d\delta_s \tag{9.30}$$

where δ_n^c and δ_s^c are the critical separations at which σ_n and σ_s drop to zero, respectively. In the model of Ortiz and Pandolfi [9-10], the cohesive energy density takes the form

$$\Gamma_c = \int_0^{\delta_{eff}^c} \sigma_{eff} d\delta_{eff} \tag{9.31}$$

where δ_{eff}^c is the effective critical separation corresponding to vanishing effective cohesive traction.

9.5.2 Cohesive Energy Density

The cohesive energy density is a constant for Mode I fracture. For fracture of bimaterial interfaces, however, the cohesive energy density may depend on the loading phase angle. Consider a crack with a cohesive zone at the interface between two dissimilar elastic materials with the shear modulus and Poisson's ratio for the upper and lower media denoted by μ_i and ν_i ($i = 1,2$), respectively, as shown in Figure 9.10.

The J-integral is still path-independent for contours surrounding the crack tip. Consider two integration paths with one (Γ_2) along the boundary of the cohesive zone and the other (Γ_1) within the dominance zone of the elastic oscillatory field

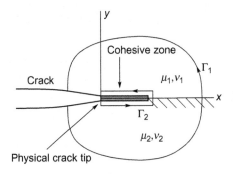

FIGURE 9.10

Integration contours around the tip of an interface crack with a cohesive zone.

(see Eqs. (8.12) and (8.13) in Chapter 8). The J-integral along Γ_1 at crack initiation is evaluated as

$$J = \frac{1}{16}\left(\frac{\kappa_1+1}{\mu_1}+\frac{\kappa_1+1}{\mu_1}\right)\left(K_1^2+K_2^2\right) = G_c$$

where G_c is the critical energy release rate. Along path Γ_2 at crack initiation, the J-integral can be evaluated as

$$J = -\int_{\Gamma_2}\sigma_{ij}n_j\frac{\partial u_i}{\partial x_1}dl = -\int_0^{L_c}\sigma_s\frac{d\delta_s}{dx}dx - \int_0^{L_c}\sigma_n\frac{d\delta_n}{dx}dx$$

$$= \int_0^{\delta_n^c}\sigma_n d\delta_n + \int_0^{\delta_s^c}\sigma_s d\delta_s = \Gamma_c$$

It follows from the path-independence of J that

$$\Gamma_c = G_c$$

Here it is assumed that both opening and sliding separations reach their critical values at the physical crack tip at crack initiation.

It is known from Chapter 8 that G_c depends on the phase angle for interface fracture. Hence, the cohesive energy density for interface fracture also depends on the phase angle, that is,

$$\Gamma_c = \Gamma_c(\psi)$$

where ψ is an appropriately defined phase angle, for example, in Eq. (8.62) in Chapter 8.

Consider the special case of $\beta = 0$ for which Mode I and Mode II deformations are uncoupled, where β is the Dundurs parameter defined in Chapter 8. In the phenomenological theory of interface fracture, the critical energy release rate G_c may be expressed as

$$G_c = G_{Ic}\left\{1 + \widetilde{G}(\psi,\lambda)\right\}$$

where G_{Ic} is the critical energy release rate for Mode I fracture and $\widetilde{G}(\psi,\lambda)$ is a dimensionless function of ψ and λ with λ being a free parameter calibrated by experiments. We thus have

$$\Gamma_c = G_{Ic}\left\{1 + \widetilde{G}(\psi,\lambda)\right\} \tag{9.32}$$

The commonly adopted functional forms of $\widetilde{G}$ include [9-12]

$$\widetilde{G}(\psi,\lambda) = \frac{(1-\lambda)\sin^2\psi}{1-(1-\lambda)\sin^2\psi}$$

$$\widetilde{G}(\psi,\lambda) = (1-\lambda)\tan^2\psi$$

$$\widetilde{G}(\psi,\lambda) = \tan^2[(1-\lambda)\psi]$$

We point out that the cohesive energy density consists of its Mode I and Mode II components, which contrasts with the fact that the classical Mode I and Mode II energy release rates of interface fracture do not exist for nonzero oscillation index ϵ as described in Chapter 8.

9.5.3 Cohesive Zone Length

In Mode I fracture, the cohesive zone is determined from the condition that no energy dissipation occurs, or the stress singularity is removed, at the cohesive zone tip. For mixed mode fracture, however, both opening and sliding separations contribute to the energy dissipation. Null energy dissipation at the tip of the cohesive zone thus requires cancellation of singularities in both normal and shear stresses at the cohesive zone tip.

Consider a crack of length $2a$ at the interface between two semi-infinite dissimilar elastic media subjected to remote tension and shear, as shown in Figure 9.11. A cohesive zone of length $\rho = c - a$ develops ahead of each crack tip upon external loading. It is assumed that the cohesive tractions are constant. Consider the small-scale cohesive zone case, that is, $\rho << a$. Hence, $a \approx a + \rho$. The complex stress intensity factor at the cohesive zone tip due to the applied loads is (Eq. (8.27) in Chapter 8)

$$K_{app} = K_1^{app} + iK_2^{app} = \frac{(\sigma_{yy}^\infty + i\sigma_{xy}^\infty)(1+2i\epsilon)}{\cosh(\pi\epsilon)}\sqrt{\pi a} \tag{9.33}$$

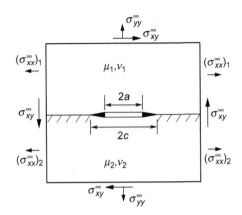

FIGURE 9.11

A crack at the interface between two dissimilar elastic materials.

and the stress intensity factor due to the cohesive traction can be obtained from the Green function solution (Eq. (8.30) in Chapter 8) as

$$K_{coh} = K_1^{coh} + iK_2^{coh} = -\sqrt{\frac{2}{\pi}}(\sigma_n + i\sigma_s)(2a)^{i\epsilon} \int_0^{\rho} \xi^{-1/2-i\epsilon} d\xi$$

$$= -\sqrt{\frac{2}{\pi}}(\sigma_n + i\sigma_s)\left(\frac{2a}{\rho}\right)^{i\epsilon} \frac{2\sqrt{\rho}}{1-2i\epsilon} \tag{9.34}$$

In these two equations, the stress intensity factors have been defined using the same length parameter $2a$ in Eq. (8.36) in Chapter 8.

The cancellation of stress singularity at the cohesive zone tip means

$$K_{app} + K_{coh} = 0 \tag{9.35}$$

or

$$Te^{i\chi}\frac{1+2i\epsilon}{\cosh(\pi\epsilon)}\sqrt{\pi a} = \sqrt{\frac{2}{\pi}}\sigma_0 e^{i\phi}\left(\frac{2a}{\rho}\right)^{i\epsilon}\frac{2\sqrt{\rho}}{1-2i\epsilon} \tag{9.36}$$

where ϕ is the phase angle of the cohesive traction and χ is the loading phase angle, that is,

$$\sigma_n + i\sigma_s = \sigma_0 e^{i\phi}$$
$$\sigma_{yy}^{\infty} + i\sigma_{xy}^{\infty} = Te^{i\chi}$$

Eq. (9.36) contains two independent conditions with only one unknown ρ. It can be satisfied only when

$$\phi = \chi + \epsilon \ln \frac{\rho}{2a} \qquad (9.37)$$

It is noted that Huang [9-13] studied the problem in the context of the Dugdale model in which ϕ is determined by Eq. (9.37). It should be pointed out that ϕ in the Dugdale model is determined as the solution of the elastic-plastic boundary value problem and ϕ thus determined may not satisfy Eq. (9.37). As a result, the stress singularity at the Dugdale zone tip may not be canceled. In the cohesive zone model, σ_n and σ_s are assumed to be material-dependent constants. While they may depend on the phase angle χ, Eq. (9.37) can not be satisfied in general, which implies that the stress singularity at the cohesive zone tip may not be canceled.

To deal with this dilemma, Jin and Sun [9-7] proposed a possible remedy that the lengths over which normal and shear tractions act are assumed to be different. For example, if we assume that ρ_n and ρ_s are the lengths of the cohesive zone segments over which normal and shear tractions act, respectively, as shown in Figure 9.12, the stress intensity factor due to the cohesive tractions would be

$$K_{coh} = -\sqrt{\frac{2}{\pi}}(2a)^{i\epsilon}\left[\sigma_n \int_0^{\rho_n} \xi^{-1/2-i\epsilon}\,d\xi + i\sigma_s \int_0^{\rho_s} \xi^{-1/2-i\epsilon}\,d\xi\right]$$

$$= -\sqrt{\frac{2}{\pi}}\left[\left(\frac{2a}{\rho_n}\right)^{i\epsilon}\frac{2\sigma_n\sqrt{\rho_n}}{1-2i\epsilon} + \left(\frac{2a}{\rho_s}\right)^{i\epsilon}\frac{2\sigma_s\sqrt{\rho_s}}{1-2i\epsilon}\right]$$

Substituting this equation and Eq. (9.33) into Eq. (9.35) yields the condition for cancelling the stress singularity:

$$\frac{(\sigma_{yy}^\infty + i\sigma_{xy}^\infty)(1+2i\epsilon)}{\cosh(\pi\epsilon)}\sqrt{\pi a} = \sqrt{\frac{2}{\pi}}\left[\left(\frac{2a}{\rho_n}\right)^{i\epsilon}\frac{2\sigma_n\sqrt{\rho_n}}{1-2i\epsilon} + \left(\frac{2a}{\rho_s}\right)^{i\epsilon}\frac{2\sigma_s\sqrt{\rho_s}}{1-2i\epsilon}\right]$$

The equation contains two independent conditions with two unknowns ρ_n and ρ_n. Hence, the stress singularity may be canceled at the cohesive zone tip.

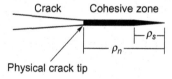

FIGURE 9.12

Different cohesive zone lengths in opening and sliding modes.

References

[9-1] G.I. Barenblatt, The mathematical theory of equilibrium cracks in brittle fracture, in: Advances in Applied Mechanics, Vol. 7, Academic Press, New York, 1962, pp. 55–129.

[9-2] Z.-H. Jin, C.T. Sun, Cohesive fracture model based on necking, Int. J. Fract. 134 (2005) 91–108.

[9-3] J.H. Rose, J. Ferrante, J.R. Smith, Universal binding energy curves for metals and bimetallic interfaces, Phys. Rev. Lett. 47 (1981) 675–678.

[9-4] G.T. Hahn, M.F. Kanninen, A.R. Rosenfield, Ductile crack extension and propagation in steel foil, in: Fracture 1969, Chapman and Hall Ltd., London, 1969, pp. 58–72.

[9-5] L.R.F. Rose, Crack reinforcement by distributed springs, J. Mech. Phys. Sol. 35 (1987) 383–405.

[9-6] R. Courant, D. Hilbert, Methods of Mathematical Physics, Interscience Publishers, New York, 1953.

[9-7] Z.-H. Jin, C.T. Sun, Cohesive zone modeling of interface fracture in elastic bimaterials, Eng. Fract. Mech. 72 (2005) 1805–1817.

[9-8] J.R. Rice, Mathematical analysis in the mechanics of fracture, in: H. Liebowitz (Ed.), Fracture, Vol. 2, Academic Press, New York, 1968, pp. 191–311.

[9-9] A. Needleman, An analysis of tensile decohesion along an interface, J. Mech. Phys. Sol. 38 (1990) 289–324.

[9-10] M. Ortiz, A. Pandolfi, Finite-deformation irreversible cohesive elements for three-dimensional crack-propagation analysis, Int. J. Numer. Methods Eng. 44 (1999) 1267–1282.

[9-11] V. Tvergaard and J.W. Hutchinson, The influence of plasticity on mixed mode interface toughness, J. Mech. Phys. Sol. 41 (1993) 1119–1135.

[9-12] J.W. Hutchinson and Z. Suo, Mixed mode cracking in layered materials, Adv. Appl. Mech. 29 (1992) 63–187.

[9-13] N.C. Huang, An estimation of the plastic zone size for a bimaterial interfacial crack, Eng. Fract. Mech. 41 (1992) 935–938.

PROBLEMS

9.1 Prove that the cohesive energy densities for the trapezoidal model Eq. (9.13) and the exponential model Eq. (9.15) are given by Eq. (9.14) and (9.16), respectively.

9.2 Find the relation between the critical separation displacements δ_c in the linear softening model Eq. (9.11) and δ_0 in the exponential model Eq. (9.15), assuming identical peak cohesive traction σ_c and identical cohesive energy density Γ_c in the two models.

9.3 Derive the integral expression of the cohesive energy density Eq. (9.31) from Eq. (9.30). Assume that the opening, sliding, and effective separations reach their critical values δ_n^c, δ_s^c, and δ_{eff}^c simultaneously.

Special Topics

This chapter briefly introduces some special topics in fracture mechanics, including fracture of anisotropic solids, fracture of nonhomegenous matertials, and dynamic fracture mechanics. In recent years, fiber-reinforced composite materials have found wide applications in aerospace, automotive, civil infrastructures, sports equipment, and other industries. Damage tolerance and defect assessments for the structural integrity of composite structures require thorough understanding of fracture behavior of composites. Fracture mechanics of nonhomogeneous materials has applications in macroscopically heterogeneous materials such as functionally graded materials (FGMs), which have been developed to meet the increasing multifunctional structural performance requirements in engineering applications. Many engineering structures are subjected to dynamic loads such as explosive loads, wind loads, impact by foreign objects, and so on. Prediction of crack initiation and propagation in a dynamically loaded structure must be based on dynamic fracture mechanics, which generally considers both inertia effects on the stresses and displacements, and loading rate effects on the constitutive responses of materials.

10.1 FRACTURE MECHANICS OF ANISOTROPIC SOLIDS

Composite materials, or simply composites, include fiber- and particulate-reinforced materials. Composites are heterogeneous materials at the microscopic scale. In continuum mechanics, properties of a composite are homogenized at the macroscopic level and the material is treated as a macroscopically, or statistically homogeneous material. The most successful composites for aerospace and other engineering applications have been fiber-reinforced polymer matrix materials. Fiber composites are usually modeled as anisotropic materials with three mutually orthogonal planes of symmerty. This section focuses on the fracture of fiber-reinforced materials. The composites are treated as anisotropic materials at the macroscopic level.

10.1.1 Basic Plane Elasticity Equations of Anisotropic Solids

The constitutive relations for a general anisotropic material are given by the matrix form

$$\{e\} = [a]\{\sigma\} \tag{10.1}$$

where $\{\sigma\}$ and $\{e\}$ are the arrays of stresses and strains defined by

$$\{\sigma\} = [\sigma_{xx}\ \sigma_{yy}\ \sigma_{zz}\ \sigma_{yz}\ \sigma_{xz}\ \sigma_{xy}]^T \tag{10.2}$$

$$\{e\} = [e_{xx}\ e_{yy}\ e_{zz}\ 2e_{yz}\ 2e_{xz}\ 2e_{xy}]^T \tag{10.3}$$

and $[a] = \{a_{ij}\}$ $(i,j = 1,2,...,6)$ is the elastic compliance matrix. If the $x - y$ plane is a plane of symmetry, then for the state of plane stress, $\sigma_z = \sigma_{yz} = \sigma_{xz} = 0$, and these constitutive equations reduce to

$$e_{xx} = a_{11}\sigma_{xx} + a_{12}\sigma_{yy} + a_{16}\sigma_{xy}$$
$$e_{yy} = a_{12}\sigma_{xx} + a_{22}\sigma_{yy} + a_{26}\sigma_{xy} \tag{10.4}$$
$$2e_{xy} = a_{16}\sigma_{xx} + a_{26}\sigma_{yy} + a_{66}\sigma_{xy}$$

with the out-of-plane normal strain given by $e_{zz} = a_{31}\sigma_{xx} + a_{32}\sigma_{yy} + a_{36}\sigma_{xy}$. The corresponding constitutive equations for the state of plane strain parallel to the $x - y$ plane are

$$e_{xx} = b_{11}\sigma_{xx} + b_{12}\sigma_{yy} + b_{16}\sigma_{xy}$$
$$e_{yy} = b_{21}\sigma_{xx} + b_{22}\sigma_{yy} + b_{26}\sigma_{xy} \tag{10.5}$$
$$2e_{xy} = b_{61}\sigma_{xx} + b_{62}\sigma_{yy} + b_{66}\sigma_{xy}$$

in which

$$b_{11} = \left(a_{11}a_{33} - a_{13}^2\right)/a_{33}, \quad b_{12} = b_{21} = (a_{12}a_{33} - a_{13}a_{23})/a_{33}$$

$$b_{22} = \left(a_{22}a_{33} - a_{23}^2\right)/a_{33}, \quad b_{16} = b_{61} = ((a_{16}a_{33} - a_{13}a_{36})/a_{33}$$

$$b_{66} = \left(a_{66}a_{33} - a_{36}^2\right)/a_{33}, \quad b_{26} = b_{62} = (a_{26}a_{33} - a_{23}a_{36})/a_{33}$$

Using the Airy stress function ϕ,

$$\sigma_{xx} = \frac{\partial^2\phi}{\partial y^2}, \quad \sigma_{xy} = -\frac{\partial^2\phi}{\partial x\partial y}, \quad \sigma_{yy} = \frac{\partial^2\phi}{\partial x^2}$$

and the plane stress constitutive relations Eq. (10.4) can be written as

$$e_{xx} = a_{11}\frac{\partial^2\phi}{\partial y^2} + a_{12}\frac{\partial^2\phi}{\partial x^2} - a_{16}\frac{\partial^2\phi}{\partial x\partial y}$$

$$e_{yy} = a_{12}\frac{\partial^2\phi}{\partial y^2} + a_{22}\frac{\partial^2\phi}{\partial x^2} - a_{26}\frac{\partial^2\phi}{\partial x\partial y}$$

$$2e_{xy} = a_{16}\frac{\partial^2\phi}{\partial y^2} + a_{26}\frac{\partial^2\phi}{\partial x^2} - a_{66}\frac{\partial^2\phi}{\partial x\partial y}$$

Substituting these equations in the following compatibility equation of strains:

$$\frac{\partial^2 e_{xx}}{\partial y^2} + \frac{\partial^2 e_{yy}}{\partial x^2} = 2\frac{\partial^2 e_{xy}}{\partial x\partial y}$$

we obtain the governing equation of the Airy stress function as follows:

$$a_{22}\frac{\partial^4\phi}{\partial x^4} - 2a_{26}\frac{\partial^4\phi}{\partial x^3\partial y} + (2a_{12}+a_{66})\frac{\partial^4\phi}{\partial x^2\partial y^2} - 2a_{16}\frac{\partial^4\phi}{\partial x\partial y^3} + a_{11}\frac{\partial^4\phi}{\partial y^4} = 0 \quad (10.6)$$

This equation may be written in the following form (Lekhnitskii [10-1]):

$$\left(\frac{\partial}{\partial y} - s_1\frac{\partial}{\partial x}\right)\left(\frac{\partial}{\partial y} - s_2\frac{\partial}{\partial x}\right)\left(\frac{\partial}{\partial y} - s_3\frac{\partial}{\partial x}\right)\left(\frac{\partial}{\partial y} - s_4\frac{\partial}{\partial x}\right)\phi = 0 \quad (10.7)$$

where the constants s_i ($i = 1,2,3,4$) are the roots of the following characteristic equation:

$$a_{11}s^4 - 2a_{16}s^3 + (2a_{12}+a_{66})s^2 - 2a_{26}s + a_{22} = 0 \quad (10.8)$$

These roots are either complex or pure imaginary according to Lekhnitskii [10-1] and can be written as

$$s_1 = \alpha_1 + i\beta_1$$
$$s_2 = \alpha_2 + i\beta_2$$
$$s_3 = \bar{s}_1 = \alpha_1 - i\beta_1$$
$$s_4 = \bar{s}_2 = \alpha_2 - i\beta_2$$

For anisotropic materials with distinct roots, the general solution of Eq. (10.7) can be expressed in terms of two complex functions $F_1(z_1)$ and $F_2(z_2)$ as follows:

$$\phi = 2\,\text{Re}[F_1(z_1) + F_2(z_2)] \quad (10.9)$$

where z_1 and z_2 are defined by

$$z_1 = x + s_1 y, \qquad z_2 = x + s_2 y$$

Hence, the stresses and displacements have the following expressions [10-1]:

$$\sigma_{xx} = 2\,\mathrm{Re}\left[s_1^2 \psi'(z_1) + s_2^2 \chi'(z_2)\right]$$

$$\sigma_{yy} = 2\,\mathrm{Re}\left[\psi'(z_1) + \chi'(z_2)\right] \qquad (10.10)$$

$$\sigma_{xy} = -2\,\mathrm{Re}\left[s_1 \psi'(z_1) + s_2 \chi'(z_2)\right]$$

$$u_x = 2\,\mathrm{Re}[p_1 \psi(z_1) + p_2 \chi(z_2)]$$

$$\qquad (10.11)$$

$$u_y = 2\,\mathrm{Re}[q_1 \psi(z_1) + q_2 \chi(z_2)]$$

where

$$\psi(z_1) = F_1'(z_1), \qquad \chi(z_2) = F_2'(z_2)$$

$$p_1 = a_{11}s_1^2 + a_{12} - a_{16}s_1, \qquad p_2 = a_{11}s_2^2 + a_{12} - a_{16}s_2$$

$$q_1 = a_{12}s_1 + a_{22}/s_1 - a_{26}, \qquad q_2 = a_{12}s_2 + a_{22}/s_2 - a_{26}$$

10.1.2 A Mode I Crack in an Infinite Anisotropic Plate under Uniform Crack Surface Pressure

Sih and Liebowitz [10-2] studied a Mode I crack of length $2a$ in an infinite anisotropic plate (see Figure 10.1) subjected to a uniform pressure σ_0 on the crack surfaces based on Lekhnitskii's complex potential formulations described earlier. The material properties need to be symmetrical about the crack line. In other words, the $x - y$ and $x - z$

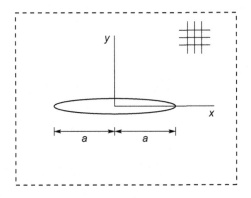

FIGURE 10.1

A crack in an infinite anisotropic plate.

planes are two mutually orthogonal planes of symmetry. Sih and Liebowitz [10-2] obtained the following complex potentials for the crack problem:

$$\psi'(z_1) = -\frac{s_2}{s_1 - s_2} \frac{\sigma_0}{2\sqrt{z_1^2 - a^2}} \left[z_1 - \sqrt{z_1^2 - a^2} \right]$$

$$\chi'(z_2) = \frac{s_1}{s_1 - s_2} \frac{\sigma_0}{2\sqrt{z_2^2 - a^2}} \left[z_2 - \sqrt{z_2^2 - a^2} \right]$$

(10.12)

The stresses and displacements near the crack tip ($x = a$, $y = 0$) can be obtained by substituting the preceding potentials into Eqs. (10.10) and (10.11), respectively, as follows

$$\sigma_{xx} = \frac{K_I}{\sqrt{2\pi r}} \text{Re} \left[\frac{s_1 s_2}{s_1 - s_2} \left(\frac{s_2}{\sqrt{\cos\theta + s_2 \sin\theta}} - \frac{s_1}{\sqrt{\cos\theta + s_1 \sin\theta}} \right) \right]$$

$$\sigma_{yy} = \frac{K_I}{\sqrt{2\pi r}} \text{Re} \left[\frac{1}{s_1 - s_2} \left(\frac{s_1}{\sqrt{\cos\theta + s_2 \sin\theta}} - \frac{s_2}{\sqrt{\cos\theta + s_1 \sin\theta}} \right) \right]$$

(10.13)

$$\sigma_{xy} = \frac{K_I}{\sqrt{2\pi r}} \text{Re} \left[\frac{s_1 s_2}{s_1 - s_2} \left(\frac{1}{\sqrt{\cos\theta + s_1 \sin\theta}} - \frac{1}{\sqrt{\cos\theta + s_2 \sin\theta}} \right) \right]$$

$$u_x = K_I \sqrt{\frac{2r}{\pi}} \text{Re} \left[\frac{1}{s_1 - s_2} \left(s_1 p_2 \sqrt{\cos\theta + s_2 \sin\theta} - s_2 p_1 \sqrt{\cos\theta + s_1 \sin\theta} \right) \right]$$

(10.14)

$$u_y = K_I \sqrt{\frac{2r}{\pi}} \text{Re} \left[\frac{1}{s_1 - s_2} \left(s_1 q_2 \sqrt{\cos\theta + s_2 \sin\theta} - s_2 q_1 \sqrt{\cos\theta + s_1 \sin\theta} \right) \right]$$

where $K_I = \sigma_0 \sqrt{\pi a}$ is the Mode I stress intensity factor (SIF), and (r, θ) are the polar coordinates centered at the crack tip. Clearly, the stresses still have the inverse square root singularity and the SIF is the same as that for an isotropic material in this special case. The angular distributions of the stresses, however, depend on the elastic constants via s_1 and s_2.

10.1.3 A Mode II Crack in an Infinite Anisotropic Plate under Uniform Crack Surface Shear

Sih and Liebowitz [10-2] also considered a Mode II crack of length $2a$ in an infinite anisotropic plate subjected to uniform crack surface shearing τ_0. The material properties are still symmetrical about the crack line. The complex potentials for the crack problem are given by

$$\psi'(z_1) = -\frac{1}{s_1 - s_2} \frac{\tau_0}{2\sqrt{z_1^2 - a^2}} \left[z_1 - \sqrt{z_1^2 - a^2} \right]$$

$$\chi'(z_2) = \frac{1}{s_1 - s_2} \frac{\tau_0}{2\sqrt{z_2^2 - a^2}} \left[z_2 - \sqrt{z_2^2 - a^2} \right]$$

(10.15)

Substituting these potentials into Eqs. (10.10) and (10.11), one can obtain the stresses and displacements near the crack tip ($x = a$, $y = 0$) as follows [10-2]:

$$\sigma_{xx} = \frac{K_{II}}{\sqrt{2\pi r}} \text{Re} \left[\frac{1}{s_1 - s_2} \left(\frac{s_2^2}{\sqrt{\cos\theta + s_2 \sin\theta}} - \frac{s_1^2}{\sqrt{\cos\theta + s_1 \sin\theta}} \right) \right]$$

$$\sigma_{yy} = \frac{K_{II}}{\sqrt{2\pi r}} \text{Re} \left[\frac{1}{s_1 - s_2} \left(\frac{1}{\sqrt{\cos\theta + s_2 \sin\theta}} - \frac{1}{\sqrt{\cos\theta + s_1 \sin\theta}} \right) \right] \quad (10.16)$$

$$\sigma_{xy} = \frac{K_{II}}{\sqrt{2\pi r}} \text{Re} \left[\frac{1}{s_1 - s_2} \left(\frac{s_1}{\sqrt{\cos\theta + s_1 \sin\theta}} - \frac{s_2}{\sqrt{\cos\theta + s_2 \sin\theta}} \right) \right]$$

$$u_x = K_{II} \sqrt{\frac{2r}{\pi}} \text{Re} \left[\frac{1}{s_1 - s_2} \left(p_2\sqrt{\cos\theta + s_2 \sin\theta} - p_1\sqrt{\cos\theta + s_1 \sin\theta} \right) \right]$$

$$u_y = K_{II} \sqrt{\frac{2r}{\pi}} \text{Re} \left[\frac{1}{s_1 - s_2} \left(q_2\sqrt{\cos\theta + s_2 \sin\theta} - q_1\sqrt{\cos\theta + s_1 \sin\theta} \right) \right]$$

(10.17)

where $K_{II} = \tau_0\sqrt{\pi a}$ is the Mode II SIF.

10.1.4 Energy Release Rate

Using the crack closure integral technique, the Mode I energy release rate may be determined from

$$G_I = \lim_{\Delta a \to 0} \frac{1}{\Delta a} \int_0^{\Delta a} \sigma_{yy}(x,0)u_y(\Delta a - x,\pi)dx \quad (10.18)$$

where $\sigma_{yy}(x,0)$ is the stress when the crack tip is at $x = 0$, and $u_y(\Delta a - x, \pi)$ is the displacement of the upper crack face when the crack tip is at $x = \Delta a$. Substituting the crack tip stress Eq. (10.13) and displacement Eq. (10.14) in the preceding equation, the energy release rates can be evaluated as follows:

$$G_I = -\frac{K_I^2}{2}a_{22} \text{Im} \left[\frac{s_1 + s_2}{s_1 s_2} \right] \quad (10.19)$$

Similarly, the Mode II energy release rate can be derived as

$$G_{II} = \lim_{\Delta a \to 0} \frac{1}{\Delta a} \int_0^{\Delta a} \sigma_{xy}(x,0) u_x(\Delta a - x, \pi) dx = \frac{K_{II}^2}{2} a_{11} \, \text{Im} \, [s_1 + s_2] \qquad (10.20)$$

For the special case of orthotropic materials with the $x-z$ and $y-z$ planes as symmetry planes, $a_{16} = a_{26} = 0$ and the roots s_1 and s_2 become

$$s_1^2 = -\frac{2a_{12} + a_{66}}{2a_{11}} + \frac{1}{2a_{11}} \sqrt{(2a_{12} + a_{66})^2 - 4a_{11}a_{22}}$$

$$s_2^2 = -\frac{2a_{12} + a_{66}}{2a_{11}} - \frac{1}{2a_{11}} \sqrt{(2a_{12} + a_{66})^2 - 4a_{11}a_{22}}$$

The energy release rates now reduce to

$$G_I = K_I^2 \sqrt{\frac{a_{11}a_{22}}{2}} \sqrt{\left[\frac{2a_{12} + a_{66}}{2a_{11}} + \sqrt{\frac{a_{22}}{a_{11}}} \right]}$$

$$G_{II} = K_{II}^2 \frac{a_{11}}{\sqrt{2}} \sqrt{\left[\frac{2a_{12} + a_{66}}{2a_{11}} + \sqrt{\frac{a_{22}}{a_{11}}} \right]}$$

$$(10.21)$$

The elastic compliances are related to the engineering moduli of the composite as

$$a_{11} = \frac{1}{E_x}, \qquad a_{22} = \frac{1}{E_y}, \qquad a_{12} = -\frac{\nu_{yx}}{E_y}, \qquad a_{66} = \frac{1}{\mu_{xy}} \qquad (10.22)$$

where E_x and E_y are the elastic moduli in the x- and y-directions, respectively, and μ_{xy} and ν_{yx} are the shear modulus and Poisson's ratio in the $x-y$ plane, respectively.

10.2 FRACTURE MECHANICS OF NONHOMOGENEOUS MATERIALS

Conventional composite materials are macroscopically, or statistically homogeneous materials. In a macroscopically nonhomogeneous material, or simply nonhomogeneous material, material properties vary with spatial position. Natural materials such as bones and bamboos are nonhomogeneous materials. Engineering materials in a field of elevated nonuniform temperature are also nonhomogeneous as material properties are temperature-dependent.

Recent progress in fracture mechanics of nonhomogeneous materials is largely owed to the development of FGMs. Broadly speaking, FGMs are materials with graded microstructures and macroproperties. For high-performance structural applications, FGMs are often multiphased materials with the volume fractions of their constituents varied gradually in predetermined profiles. This chapter introduces some basic concepts of fracture mechanics of nonhomogeous elastic materials, including FGMs.

10.2.1 Basic Plane Elasticity Equations of Nonhomogenous Materials

The basic equations of nonhomogeneous materials have the same forms as those for homogeneous solids except that the material properties are functions of spatial position. In plane elasticity of nonhomogeneous materials, Hooke's law with spatial position-dependent elasticity coefficients can be written as

$$
\begin{aligned}
\sigma_{xx} &= \lambda^*(x,y)\left(e_{xx}+e_{yy}\right)+2\mu(x,y)e_{xx} \\
\sigma_{yy} &= \lambda^*(x,y)\left(e_{xx}+e_{yy}\right)+2\mu(x,y)e_{yy} \\
\sigma_{xy} &= 2\mu(x,y)e_{xy}
\end{aligned}
\tag{10.23}
$$

where $\lambda(x,y)$ and $\mu(x,y)$ are the spatial position-dependent Lame's constants, which are related to Young's modulus $E(x,y)$ and Poisson's ratio $v(x,y)$ by

$$
\lambda(x,y) = \frac{E(x,y)v(x,y)}{[1+v(x,y)][1-2v(x,y)]}, \qquad \mu(x,y) = \frac{E(x,y)}{2[1+v(x,y)]}
$$

and $\lambda^*(x,y)$ is given by

$$
\lambda^* = \begin{cases} \lambda & \text{for plane strain} \\ \dfrac{2\lambda\mu}{\lambda+2\mu} & \text{for plane stress} \end{cases}
$$

The inverse form of Eq. (10.23) can be expressed as follows:

$$
\begin{aligned}
e_{xx} &= \frac{1}{2\mu(x,y)}\left[\sigma_{xx} - \frac{\lambda^*(x,y)}{2\left(\lambda^*(x,y)+\mu(x,y)\right)}\left(\sigma_{xx}+\sigma_{yy}\right)\right] \\
e_{yy} &= \frac{1}{2\mu(x,y)}\left[\sigma_{yy} - \frac{\lambda^*(x,y)}{2\left(\lambda^*(x,y)+\mu(x,y)\right)}\left(\sigma_{xx}+\sigma_{yy}\right)\right] \\
e_{xy} &= \frac{1}{2\mu(x,y)}\sigma_{xy}
\end{aligned}
\tag{10.24}
$$

In the continuum analysis of graded composites, the elastic properties can be calculated from a micromechanics model (e.g., the Mori-Tanaka model derived by Weng [10-3]), or can be assumed to follow some elementary functions that are consistent with micromechanics analyses. An exponentially varying Young's modulus and a constant Poisson's ratio are frequently used, which leads to constant coefficient, partial differential equations for the displacement fields.

In addition to the constitutive law, the equilibrium equations,

$$\frac{\partial \sigma_{xx}}{\partial x} + \frac{\partial \sigma_{xy}}{\partial y} = 0$$

$$\frac{\partial \sigma_{xy}}{\partial x} + \frac{\partial \sigma_{yy}}{\partial y} = 0$$

(10.25)

and the strain-displacement relations,

$$e_{xx} = \frac{\partial u_x}{\partial x}, \quad e_{yy} = \frac{\partial u_y}{\partial y}, \quad e_{xy} = \frac{1}{2}\left(\frac{\partial u_x}{\partial y} + \frac{\partial u_y}{\partial x}\right)$$

(10.26)

should also be satisfied.

By eliminating stresses and strains in Eqs. (10.23), (10.25), and (10.26), we can obtain the following governing equations for the displacements u_x and u_y:

$$(\lambda^* + \mu)\frac{\partial}{\partial x}\left(\frac{\partial u_x}{\partial x} + \frac{\partial u_y}{\partial y}\right) + \mu \nabla^2 u_x$$

$$+ \frac{\partial \lambda^*}{\partial x}\left(\frac{\partial u_x}{\partial x} + \frac{\partial u_y}{\partial y}\right) + 2\frac{\partial \mu}{\partial x}\frac{\partial u_x}{\partial x} + \frac{\partial \mu}{\partial y}\left(\frac{\partial u_x}{\partial y} + \frac{\partial u_y}{\partial x}\right) = 0,$$

$$(\lambda^* + \mu)\frac{\partial}{\partial y}\left(\frac{\partial u_x}{\partial x} + \frac{\partial u_y}{\partial y}\right) + \mu \nabla^2 u_y$$

$$+ \frac{\partial \lambda^*}{\partial y}\left(\frac{\partial u_x}{\partial x} + \frac{\partial u_y}{\partial y}\right) + 2\frac{\partial \mu}{\partial y}\frac{\partial u_y}{\partial y} + \frac{\partial \mu}{\partial x}\left(\frac{\partial u_x}{\partial y} + \frac{\partial u_y}{\partial x}\right) = 0$$

(10.27)

where ∇^2 is the Laplace operator. For a nonhomogeneous material with exponentially graded modulus and constant Poisson's ratio, that is,

$$\mu = \mu_0 \exp(\beta x + \gamma y), \qquad \nu = \nu_0$$

(10.28)

where μ_0, ν_0, β, and γ are material constants, the governing equations (10.27) for the displacements reduce to

$$(\kappa + 1)\frac{\partial^2 u_x}{\partial x^2} + (\kappa - 1)\frac{\partial^2 u_x}{\partial y^2} + 2\frac{\partial u_y}{\partial x \partial y} + \beta(\kappa + 1)\frac{\partial u_x}{\partial x}$$

$$+ \gamma(\kappa - 1)\frac{\partial u_x}{\partial y} + \gamma(\kappa - 1)\frac{\partial u_y}{\partial x} + \beta(3 - \kappa)\frac{\partial u_y}{\partial y} = 0,$$

$$(\kappa - 1)\frac{\partial^2 u_y}{\partial x^2} + (\kappa + 1)\frac{\partial^2 u_y}{\partial y^2} + 2\frac{\partial u_x}{\partial x \partial y} + \gamma(3 - \kappa)\frac{\partial u_x}{\partial x}$$

$$+ \beta(\kappa - 1)\frac{\partial u_x}{\partial y} + \beta(\kappa - 1)\frac{\partial u_y}{\partial x} + \gamma(\kappa + 1)\frac{\partial u_y}{\partial y} = 0$$

(10.29)

where $\kappa = 3 - 4\nu_0$ for plane strain, and $\kappa = (3 - \nu_0)/(1 + \nu_0)$ for plane stress. The equations in Eq. (10.29) are constant coefficient, partial differential equations that could be treated analytically or semi-analytically.

Equation (10.27) is used in the displacement method for plane elasticity of nonhomogeneous materials. Equivalently, a stress function method can be adopted. In this case, stresses are expressed in terms of the Airy stress function ϕ as follows:

$$\sigma_{xx} = \frac{\partial^2 \phi}{\partial y^2}, \qquad \sigma_{yy} = \frac{\partial^2 \phi}{\partial x^2}, \qquad \sigma_{xy} = -\frac{\partial^2 \phi}{\partial x \partial y} \tag{10.30}$$

The equlibrium equations (10.25) are satisfied with these stresses. Use of Hooke's law Eq. (10.24) and the strain-displacement relations Eq. (10.26) yields the governing equation of the Airy stress function for general nonhomogeneous materials under plane stress conditions:

$$\frac{1}{E} \nabla^2 \nabla^2 \phi + 2 \left[\frac{\partial}{\partial x} \left(\frac{1}{E} \right) \frac{\partial \nabla^2 \phi}{\partial x} + \frac{\partial}{\partial y} \left(\frac{1}{E} \right) \frac{\partial \nabla^2 \phi}{\partial y} \right] + \nabla^2 \left(\frac{1}{E} \right) \nabla^2 \phi$$

$$- \frac{\partial^2}{\partial y^2} \left(\frac{1+\nu}{E} \right) \frac{\partial^2 \phi}{\partial x^2} - \frac{\partial^2}{\partial x^2} \left(\frac{1+\nu}{E} \right) \frac{\partial^2 \phi}{\partial y^2} + 2 \frac{\partial^2}{\partial x \partial y} \left(\frac{1+\nu}{E} \right) \frac{\partial^2 \phi}{\partial x \partial y} = 0 \tag{10.31}$$

For plane strain, E and ν in this equation should be replaced by $E/(1 - \nu^2)$ and $\nu/(1 - \nu)$, respectively. Again, consider the nonhomogeneous material with a constant Poisson's ratio and an exponentially varying modulus Eq. (10.28). Now the basic equation of the Airy stress function Eq. (10.31) reduces to

$$\nabla^2 \nabla^2 \phi - 2 \left(\beta \frac{\partial}{\partial x} + \gamma \frac{\partial}{\partial y} \right) \nabla^2 \phi + \left(\beta^2 - \nu_0 \gamma^2 \right) \frac{\partial^2 \phi}{\partial x^2}$$

$$+ 2(1 + \nu_0) \beta \gamma \frac{\partial^2 \phi}{\partial x \partial y} + \left(\gamma^2 - \nu_0 \beta^2 \right) \frac{\partial^2 \phi}{\partial y^2} = 0 \tag{10.32}$$

10.2.2 Crack Tip Stress and Displacement Fields

Consider a cracked nonhomogeneous material with continuous and piecewise differentiable Young's modulus E and Poisson's ratio ν. Here we assume there are only two differentiable pieces as shown in Figure 10.2, but the conclusion for the crack tip fields hold true for general cases. The boundary between the two differentiable pieces is referred to as the "weak property discontinuity line" (line L as shown in Figure 10.2). E and ν are continuously differentiable in each piece, continuous across the boundary L, and the derivatives of E and ν with respect to the spatial coordinates may undergo jumps across L. Assume that the crack terminates at the boundary L at an angle θ_L. The crack becomes an interface crack when θ_L equals 0 or π.

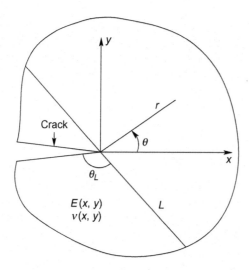

FIGURE 10.2

A crack in a nonhomogeneous medium.

To obtain the crack tip stress and deformation fields, assume the following asymptotic expansion of the Airy stress function in each differentiable piece near the crack tip (Jin and Noda [10-4]):

$$\phi = r^{s_1} \tilde{F}_1(\theta), \quad r \to 0, \quad -\pi < \theta < -(\pi - \theta_L),$$

$$\phi = r^{s_2} \tilde{F}_2(\theta), \quad r \to 0, \quad -(\pi - \theta_L) < \theta < \theta_L, \qquad (10.33)$$

$$\phi = r^{s_3} \tilde{F}_3(\theta), \quad r \to 0, \quad \theta_L < \theta < \pi$$

where (r, θ) are the polar coordinates centered at the crack tip with $\theta = \pm\pi$ describing the crack faces, s_i $(i = 1, 2, 3)$ the eigenvalues to be determined, and $\tilde{F}_i$ $(i = 1, 2, 3)$ the unknown functions of θ. As the material is assumed to be piecewise differentiable, the elastic properties can be expanded into a Taylor series at the crack tip in each differentiable piece, that is,

$$E = E_{tip} + a_{11}x + b_{11}y + c_{11}x^2 + d_{11}xy + e_{11}y^2 + ..., \quad -\pi < \theta < -(\pi - \theta_L),$$

$$E = E_{tip} + a_{12}x + b_{12}y + c_{12}x^2 + d_{12}xy + e_{12}y^2 + ..., \quad -(\pi - \theta_L) < \theta < \theta_L,$$

$$E = E_{tip} + a_{13}x + b_{13}y + c_{13}x^2 + d_{13}xy + e_{13}y^2 + ..., \quad \theta_L < \theta < \pi$$

$$(10.34)$$

where $(x, y) = r(\cos\theta, \sin\theta)$ are the rectangular coordinates at the crack tip, E_{tip} is the Young's modulus at the crack tip, and $a_{ij}, b_{ij}, c_{ij}, d_{ij}$, and e_{ij} are constants related to the derivatives of the modulus at the crack tip in the differential pieces.

Similarly, the Poisson's ratio can also be expanded into a Taylor series in each differentiable piece. By substituting Eqs. (10.33) and (10.34) into the governing equation (10.31), we can find that the first term dominates the other terms, and hence the dominant term for the Airy stress function still satisfies the biharmonic equation in every differentiable piece, that is,

$$\nabla^2 \nabla^2 \phi = 0 \tag{10.35}$$

The singular solution to homogeneous materials (Williams [10-5]) satisfies the same equation. Thus, it is also the dominant solution to the nonhomogeneous material in every differentiable piece and satisfies the displacement and traction continuity conditions across the weak property discontinuity line as long as the material properties are continuous. Hence, we can come to the conclusion that the crack tip stress and displacement fields in nonhomogeneous materials have the same forms as those in homogeneous materials provided the material properties are continuous and piecewise differentiable. The elastic properties at the crack tip, however, should be used in the displacement field.

Eischen [10-6] considered a nonhomogeneous material with continuously differentiable properties and reached the same conclusion. Thus, the asymptotic crack tip stress and displacement fields in nonhomogeneous elastic materials can be written as $(r \to 0)$

$$
\begin{aligned}
\sigma_{xx} &= \frac{K_I}{\sqrt{2\pi r}} \cos\frac{1}{2}\theta \left(1 - \sin\frac{1}{2}\theta \sin\frac{3}{2}\theta\right) \\
&\quad - \frac{K_{II}}{\sqrt{2\pi r}} \sin\frac{1}{2}\theta \left(2 + \cos\frac{\theta}{2}\cos\frac{3\theta}{2}\right) \\
\sigma_{yy} &= \frac{K_I}{\sqrt{2\pi r}} \cos\frac{1}{2}\theta \left(1 + \sin\frac{1}{2}\theta \sin\frac{3}{2}\theta\right) \\
&\quad + \frac{K_{II}}{\sqrt{2\pi r}} \sin\frac{\theta}{2}\cos\frac{\theta}{2}\cos\frac{3}{2}\theta \\
\sigma_{xy} &= \frac{K_I}{\sqrt{2\pi r}} \sin\frac{1}{2}\theta \cos\frac{1}{2}\theta \cos\frac{3}{2}\theta \\
&\quad + \frac{K_{II}}{\sqrt{2\pi r}} \cos\frac{1}{2}\theta \left(1 - \sin\frac{1}{2}\theta \sin\frac{3}{2}\theta\right)
\end{aligned}
\tag{10.36}
$$

and

$$
\begin{aligned}
u_x &= \frac{K_I}{8\mu_{tip}\pi} \sqrt{2\pi r} \left[(2\kappa_{tip} - 1)\cos\frac{\theta}{2} - \cos\frac{3\theta}{2} \right] \\
&\quad + \frac{K_{II}}{8\mu_{tip}\pi} \sqrt{2\pi r} \left[(2\kappa_{tip} + 3)\sin\frac{\theta}{2} + \sin\frac{3\theta}{2} \right]
\end{aligned}
$$

$$u_y = \frac{K_I}{8\mu_{tip}\pi}\sqrt{2\pi r}\left[(2\kappa_{tip}+1)\sin\frac{\theta}{2}-\sin\frac{3\theta}{2}\right]$$

$$-\frac{K_{II}}{8\mu_{tip}\pi}\sqrt{2\pi r}\left[(2\kappa_{tip}-3)\cos\frac{\theta}{2}+\cos\frac{3\theta}{2}\right] \qquad (10.37)$$

where K_I and K_{II} are Mode I and Mode II SIFs, respectively, and μ_{tip} and ν_{tip} are the shear modulus and Poisson's ratio at the crack tip, and κ_{tip} is given by

$$\kappa_{tip} = \begin{cases} 3-4\nu_{tip} & \text{for plane strain} \\ \dfrac{3-\nu_{tip}}{1+\nu_{tip}} & \text{for plane stress} \end{cases}$$

The previous result about the identity of the crack tip fields between homogeneous and nonhomogeneous materials is obtained under two-dimensional quasistatic, isotropic conditions. The conclusion holds true for general three-dimensional, anisotropic, and dynamic crack problems of nonhomogeneous materials. Equations (10.36) and (10.37) indicate that material nonhomogeneities influence the crack tip stress and displacement solutions only through SIFs. Because the crack tip stress and displacement fields have the same forms as those for homogeneous materials, the relation between the energy release rate G and SIFs for nonhomogeneous materials also have the form

$$G = \frac{1}{E^*_{tip}}\left(K_I^2 + K_{II}^2\right)$$

where $E^*_{tip} = E_{tip}$ for plane stress and $E^*_{tip} = E_{tip}/(1-\nu^2_{tip})$ for plane strain.

The asymptotic stress and displacement fields Eqs. (10.36) and (10.37) are only valid at points very close to the crack tip as compared with the crack length or any other in-plane characteristic lengths of the cracked body. While gradients of the elastic moduli do not influence the inverse square-root singularity, they may affect the size of the K-dominance zone in which the solution Eq. (10.36) holds. A simple estimate of the effect of material gradation on the size of the K-dominance zone may be made on the basis of Eq. (10.31) and the asymptotic solution Eq. (10.36). For a Mode I crack, at a radial distance r from the crack tip,

$$\frac{\partial^2\phi}{\partial x_\alpha \partial x_\beta} \sim \frac{K_I}{\sqrt{2\pi r}}$$

to within an angular multiplier of order unity. This singularity is due to the first term in Eq. (10.31). Neglecting gradients of Poisson's ratio, the dominance of this first term over the other terms involving modulus gradients in Eq. (10.31) leads to the K-dominance conditions related to material nonhomogeneities [10-7]:

$$\frac{1}{E}\left|\frac{\partial E}{\partial x_\alpha}\right| << \frac{1}{r}, \qquad \frac{1}{E}\left|\frac{\partial^2 E}{\partial x_\alpha \partial x_\beta}\right| << \frac{1}{r^2}. \qquad (10.38)$$

This condition shows that the size of the K-dominance zone decreases with increasing magnitude of modulus gradients. The K-dominance zone becomes vanishingly small for a crack located in a nearly sharp interface region where the modulus gradients become extremely steep.

When the modulus gradients are moderate in the crack tip region and a K-dominance zone exists, crack growth can be predicted based on the SIF criterion, that is

$$K_I = K_{Ic}^{tip}$$

where K_{Ic}^{tip} is the fracture toughness value at the crack tip location. Note that for nonhomogenous materials, fracture toughness is generally a function of spatial position, that is, $K_{Ic} = K_{Ic}(x,y)$.

10.2.3 Energy Release Rate

We know from Chapter 4 that for homogeneous elastic materials, the energy release rate is represented by a contour integral—the path-independent J-integral. This section derives the integral representation of energy release rate for nonhomogeneous materials [10-8]. It will be seen that the energy release rate can no longer be represented by a contour integral only. Similar to the approach for homogeneous materials introduced in Chapter 4, we onsider a two-dimensonal nonhomogeneous medium with a crack of length a shown shown in Figure 10.3. The area of the cracked medium is denoted by A_0 and the the boundary is Γ_0. The boundary Γ_0 consists of the outer contour Γ and the crack surfaces Γ_a, that is,

$$\Gamma_0 = \Gamma \cup \Gamma_a$$

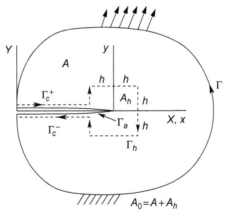

FIGURE 10.3

Contours and coordinate systems in a nonhomogeneous medium with a crack.

The medium is subjected to the prescribed traction T_i along the boundary segment Γ_t and the prescribed displacements on the boundary segment Γ_u. The crack surfaces are along the X-axis and are assumed to be free of traction. The positive contour direction of Γ_0 is defined in that when one travels along it, the domain of interest always lies to the left of the traveler. The potential energy Π of the cracked system per unit thickness can be written as

$$\Pi = \Pi(a) = \iint_{A_0} W dX dY - \int_{\Gamma_t} T_i u_i d\Gamma \qquad (10.39)$$

where a is the crack length and (X, Y) is a stationary Cartesian coordinate system. Again, the body forces are absent. For nonhomogeneous materials, the strain energy density W in Eq. (10.39) is an explicit function of spatial position and is given by

$$W = \mu(X, Y) \left[e_{ij} e_{ij} + \frac{v(X, Y)}{1 - 2v(X, Y)} (e_{kk})^2 \right] \qquad (10.40)$$

The energy release rate is defined by

$$G = -\frac{d\Pi}{da} = -\frac{d}{da} \iint_{A_0} W dX dY + \frac{d}{da} \int_{\Gamma_t} T_i u_i d\Gamma \qquad (10.41)$$

We now consider a small square A_h with the center at the crack tip and boundary Γ_h as shown in Figure 10.3. The region of the cracked body excluding A_h is denoted by A, that is,

$$A_0 = A \cup A_h$$

Because no stress singularity exists in A and along Γ_t, Eq. (10.41) can be written as

$$G = -\frac{d}{da} \left[\iint_A W dX dY + \iint_{A_h} W dX dY \right] + \int_{\Gamma_t} T_i \frac{du_i}{da} d\Gamma$$

$$= -\iint_A \frac{dW}{da} dX dY + \int_{\Gamma_0} T_i \frac{du_i}{da} d\Gamma - \frac{d}{da} \iint_{A_h} W dX dY \qquad (10.42)$$

Here the integration along Γ_t is extended to the entire boundary Γ_0 because $T_i = 0$ on the crack faces Γ_a and $du_i/da = 0$ on Γ_u. By using a local coordinate system (x, y) attached at the crack tip,

$$x = X - a, \quad y = Y$$

$$\frac{d(\ )}{da} = \frac{\partial(\ )}{\partial a} - \frac{\partial(\ )}{\partial x}$$

Equation (10.42) can be written as

$$G = -\iint_A \frac{\partial W}{\partial a} dxdy + \int_{\Gamma_0} T_i \frac{\partial u_i}{\partial a} d\Gamma$$

$$+ \iint_A \frac{\partial W}{\partial x} dxdy - \int_{\Gamma_0} T_i \frac{\partial u_i}{\partial x} d\Gamma - \frac{d}{da} \iint_{A_h} W dXdY \qquad (10.43)$$

For nonhomogeneous materials, using the divengence theorem and traction-free conditions along the crack surfaces gives

$$\iint_A \frac{\partial W}{\partial a} dxdy = \int_{\Gamma+\Gamma_h} T_i \frac{\partial u_i}{\partial a} d\Gamma + \iint_A \left(\frac{\partial W}{\partial x}\right)_{\text{expl}} dxdy$$

where $(\partial W/\partial x)_{\text{expl}}$ denotes the explicit derivative of W with respect to x, that is,

$$\left(\frac{\partial W}{\partial x}\right)_{\text{expl}} = \frac{\partial W(e_{ij}, x, y)}{\partial x}\Big|_{y=const., e_{ij}=const.}$$

Note that on the crack surface $T_i = dy = 0$. Use of the divergence theorem thus gives

$$\iint_A \frac{\partial W}{\partial x} dA = \int_{\Gamma+\Gamma_h} W n_x d\Gamma = \int_{\Gamma+\Gamma_h} W dy$$

Hence,

$$G = \int_\Gamma W dy - \int_\Gamma T_i \frac{\partial u_i}{\partial x} d\Gamma - \iint_A \left(\frac{\partial W}{\partial x}\right)_{\text{expl}} dxdy$$

$$+ \int_{\Gamma_h} W dy - \int_{\Gamma_h} T_i \frac{\partial u_i}{\partial a} d\Gamma - \frac{d}{da} \iint_{A_h} W dXdY \qquad (10.44)$$

Based on the results in the previous section, the strain energy density function for nonhomogeneous materials has the following universal separable form in the region near the moving crack tip:

$$W = B(a)\widetilde{W}(X - a, Y) = B(a)\widetilde{W}(x, y) \qquad (10.45)$$

where $B(a)$ may depend on loading and other factors but not on the local coordinates, and $\widetilde{W}(x, y)$ is a function of local coordinates only. Now assume that A_h is so small that Eq. (10.45) holds in a region containing A_h. It can be shown that the last three terms on the right-hand side of Eq. (10.44) reduce to

$$-\iint_{A_h} \left(\frac{\partial W}{\partial x}\right)_{\text{expl}} dxdy$$

We thus obtain the expression of energy release rate for nonhomogeneous materials:

$$G = \int_{\Gamma} \left(Wdy - T_i \frac{\partial u_i}{\partial x} d\Gamma\right) - \iint_{A_0} \left(\frac{\partial W}{\partial x}\right)_{\text{expl}} dxdy \qquad (10.46)$$

This expression shows that for nonhomogeneous materials, the energy release rate consists of a contour and a domain integral. The domain integral is due to the explicit dependence of strain energy density on the spatial position as shown in Eq. (10.40). The energy release rate in Eq. (10.46) is the path/domain-independent J^*-integral introduced by Eischen [10-6].

Consider a special nonhomgeneous elastic material with an exponentially graded shear modulus μ and constant Poisson's ratio ν:

$$\mu = \mu_0 \exp(\beta X), \qquad \nu = \nu_0$$

The strain energy density Eq. (10.40) for this material becomes

$$W = \mu_0 \exp(\beta X) \left[e_{ij}e_{ij} + \frac{\nu_0}{1 - 2\nu_0}(e_{kk})^2\right]$$

The explicit derivative of W with respect to x is

$$\left(\frac{\partial W}{\partial x}\right)_{\text{expl}} = \beta\mu_0\exp(\beta X)\left[e_{ij}e_{ij} + \frac{\nu_0}{1 - 2\nu_0}(e_{kk})^2\right] = \beta W$$

Note that

$$\iint_{A_0} \beta W dxdy = \frac{\beta}{2}\iint_{A_0} \sigma_{ij}u_{i,j}dxdy = \frac{\beta}{2}\int_{\Gamma+\Gamma_a} \sigma_{ij}u_in_jd\Gamma = \frac{\beta}{2}\int_{\Gamma} T_iu_id\Gamma$$

The energy release rate G in Eq. (10.46) now becomes a contour integral:

$$G = \int_{\Gamma} \left(Wdy - T_i \frac{\partial u_i}{\partial x}d\Gamma - \frac{\beta}{2}T_iu_id\Gamma\right)$$

This is the path-independent J_e-integral presented by Honein and Herrmann [10-9].

10.2.4 Stress Intensity Factors for a Crack in a Graded Interlayer between Two Dissimilar Materials

One of the advantages of graded materials is to eliminate material property mismatch at a sharp interface between two dissimilar materials thereby enhancing the interfacial bonding strength and reducing the thermal residual stresses at the interface. Delale and Erdogan [10-10] considered two dissimilar homogeneous semi-infinite media bonded through a graded interlayer with a crack parallel to the interfaces as shown in Figure 10.4, where h_0 is the thickness of the graded layer, h_1 and h_2 the distances between the crack and the upper and lower interfaces, respectively, and a the half crack length.

Delale and Erdogan [10-10] assumed the following Young's modulus and Poisson's ratio in the graded layer:

$$E = E_0 \exp(\beta y), \quad v = v_0(1 + \gamma y)\exp(\beta y), \quad -h_2 \le y \le h_1, \quad (10.47)$$

where E_0, v_0, β, and γ are material constants that can be determined using the continuity requirements of material properties across the interfaces between the graded layer and the homogeneous half planes as follows:

$$
E_0 = E_1 \exp(-\beta h_1), \quad \beta = \frac{1}{h_0} \ln \frac{E_1}{E_2},
$$

$$
v_0 = v_1 \frac{\exp(-\beta h_1)}{1 + \gamma h_1}, \quad \gamma = \frac{v_1 - v_2 \exp(\beta h_0)}{v_1 h_2 + v_2 h_1 \exp(\beta h_0)}, \quad (10.48)
$$

where E_1 and v_1 are the Young's modulus and Poisson's ratio of the upper homogeneous half plane ($y \ge h_1$), respectively, and E_2 and v_2 the material properties of the lower homogeneous half plane ($y \le -h_2$). The function $v(y)$ given in

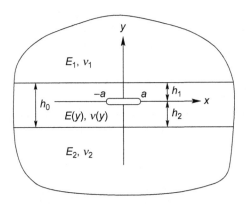

FIGURE 10.4

Two dissimilar homogeneous materials bonded by a graded interlayer with a crack parallel to the interfaces.

Eq. (10.48) should satisfy the physical requirement of $-1 \le v \le 0.5$. The governing equation (10.31) in the graded layer now reduces to

$$\nabla^2\nabla^2\phi - 2\beta\frac{\partial}{\partial y}\nabla^2\phi + \beta^2\frac{\partial^2\phi}{\partial y^2} = 0, \quad -h_2 < y < h_1 \qquad (10.49)$$

In the homogeneous half planes, the Airy stress function satisfies the biharmonic equation, that is,

$$\nabla^2\nabla^2\phi = 0, \quad y > h_1 \text{ and } y < -h_2$$

Delale and Erdogan [10-10] considered the following material combination: $E_1 = 3 \times 10^4$ ksi = 206.85 GPa, $v_1 = 0.3$, and $E_2 = 10^4$ ksi = 68.95 GPa, $v_2 = 0.3$. Table 10.1 lists SIFs (normalized by $p_0\sqrt{\pi a}$) for various values of h_1/h_0 and a/h_1 under uniform pressure load p_0 on the crack faces, and Table 10.2 shows the SIFs (normalized by $q_0\sqrt{\pi a}$) under the uniform shearing load q_0 on the crack faces. More detailed results can be found in Delale and Erdogan [10-10]. Unlike the oscillatory nature of stress and displacement fields found in the sharp interface crack problems (Chapter 8), the usual inverse square-root singularity now prevails and the SIF is well defined for the crack in this composite system due to the continuous variations of material properties.

It is seen from the tables that under uniform pressure loading, both the normalized Mode I and Mode II SIFs increase with increasing a/h_1 for a given crack location in the graded layer when $h_1/h_0 \le 0.5$. For a given crack length, the Mode I SIF increases with decreasing h_1/h_0 indicating material gradation effects. The Mode I SIFs are negative under uniform shear loading, indicating crack face contact in the crack tip region only if the shearing load is applied at the crack faces.

Table 10.1 Normalized Stress Intensity Factors under Uniform Crack Face Pressure p_0

h_1/h_0	a/h_1	$K_I/(p_0\sqrt{\pi a})$	$K_{II}/(p_0\sqrt{\pi a})$
	0.1	1.001	0.001
0.25	1.0	1.044	0.051
	8.0	1.159	0.164
	0.1	1.001	0.004
0.50	1.0	1.026	0.092
	8.0	1.046	0.165
	0.1	0.998	0.011
0.75	1.0	0.950	0.098
	8.0	0.915	0.158

Source: Adapted from Delale and Erdogan [10-10].

Table 10.2 Normalized Stress Intensity Factors under Uniform Crack Face Shear q_0

h_1/h_0	a/h_1	$K_I/(q_0\sqrt{\pi a})$	$K_{II}/(q_0\sqrt{\pi a})$
	0.1	−0.001	1.000
0.25	1.0	−0.060	1.011
	8.0	−0.152	1.072
	0.1	−0.004	1.000
0.50	1.0	−0.089	1.002
	8.0	−0.162	1.024
	0.1	−0.011	0.998
0.75	1.0	−0.101	0.973
	8.0	−0.160	0.926

Source: Adapted from Delale and Erdogan [10-10].

10.3 DYNAMIC FRACTURE MECHANICS

Many engineering structures are designed to resist dynamic loads such as explosive loads, wind loads, impact by foreign objects, and so on. The behavior of cracks in a dynamically loaded structure differs from that under quasistatic loads. Moreover, material properties also become rate-dependent when the loading rate is high. Dynamic fracture mechanics deals with crack initiation and propagation in materials and structures when the inertia effects can not be ignored.

Generally speaking, there are two kinds of dynamic fracture problems, that is, stationary cracks under dynamic loading and rapid crack propagation and crack arrest. In this chapter, we introduce some basic concepts and theories of dynamic fracture mechanics in the framework of linear elasticity. The rate effects of constitutive relations of materials thus will not be a concern. However, the rate dependence of dynamic fracture toughness will be included in the fracture criteria.

10.3.1 Basic Equations of Plane Elastodynamics

The basic equations of plane elastodynamics consist of equations of motion,

$$\frac{\partial \sigma_{xx}}{\partial x} + \frac{\partial \sigma_{xy}}{\partial y} = \rho \ddot{u}_x$$

$$\frac{\partial \sigma_{xy}}{\partial x} + \frac{\partial \sigma_{yy}}{\partial y} = \rho \ddot{u}_y$$

(10.50)

strain-displacement relations,

$$e_{xx} = \frac{\partial u_x}{\partial x}, \quad e_{yy} = \frac{\partial u_y}{\partial y}, \quad e_{xy} = \frac{1}{2}\left(\frac{\partial u_x}{\partial y} + \frac{\partial u_y}{\partial x}\right)$$

(10.51)

and Hooke's law,

$$\sigma_{xx} = \lambda^* \left(e_{xx} + e_{yy}\right) + 2\mu e_{xx}$$
$$\sigma_{yy} = \lambda^* \left(e_{xx} + e_{yy}\right) + 2\mu e_{yy} \qquad (10.52)$$
$$\sigma_{xy} = 2\mu e_{xy}$$

In these equations, a dot over a quantity denotes its material derivative with respect to time, t is time, ρ is the mass density, λ and μ are Lame constants related to Young's modulus and Poisson's ratio by

$$\lambda = \frac{E\nu}{(1+\nu)(1-2\nu)}, \qquad \mu = \frac{E}{2(1+\nu)} \qquad (10.53)$$

and λ^* is given by

$$\lambda^* = \begin{cases} \lambda & \text{for plane strain} \\ \dfrac{2\lambda\mu}{\lambda+2\mu} & \text{for plane stress} \end{cases} \qquad (10.54)$$

The stresses can be expressed in terms of displacements using Hooke's law and strain-displacement relations as follows:

$$\sigma_{xx} = \lambda^* \left(\frac{\partial u_x}{\partial x} + \frac{\partial u_y}{\partial y}\right) + 2\mu \frac{\partial u_x}{\partial x}$$
$$\sigma_{yy} = \lambda^* \left(\frac{\partial u_x}{\partial x} + \frac{\partial u_y}{\partial y}\right) + 2\mu \frac{\partial u_y}{\partial y} \qquad (10.55)$$
$$\sigma_{xy} = \mu \left(\frac{\partial u_x}{\partial y} + \frac{\partial u_y}{\partial x}\right)$$

Substituting these stresses into the equations of motion yields the following governing equations of displacements:

$$\mu\nabla^2 u_x + (\lambda^* + \mu)\frac{\partial}{\partial x}\left(\frac{\partial u_x}{\partial x} + \frac{\partial u_y}{\partial y}\right) = \rho\ddot{u}_x$$
$$\mu\nabla^2 u_y + (\lambda^* + \mu)\frac{\partial}{\partial y}\left(\frac{\partial u_x}{\partial x} + \frac{\partial u_y}{\partial y}\right) = \rho\ddot{u}_y \qquad (10.56)$$

The basic equations here can be converted to the standard wave equations using the following displacement potentials φ and ψ:

$$u_x = \frac{\partial\varphi}{\partial x} + \frac{\partial\psi}{\partial y}, \qquad u_y = \frac{\partial\varphi}{\partial y} - \frac{\partial\psi}{\partial x} \qquad (10.57)$$

Substituting this equation into Eq. (10.56) leads to the following wave equations for the potentials φ and ψ:

$$\nabla^2 \varphi = \frac{1}{c_1^2} \ddot{\varphi}, \qquad \nabla^2 \psi = \frac{1}{c_2^2} \ddot{\psi} \qquad (10.58)$$

where ∇^2 is the Laplace operator

$$\nabla^2 = \frac{\partial^2}{\partial x^2} + \frac{\partial^2}{\partial y^2}$$

and c_1 and c_2 are given by

$$c_1 = \sqrt{\frac{\lambda^* + 2\mu}{\rho}}, \qquad c_2 = \sqrt{\frac{\mu}{\rho}} \qquad (10.59)$$

c_1 and c_2 are the dilatational and shear wave speeds of the medium, respectively.

10.3.2 Stationary Cracks under Dymanic Loading

To study crack initiation under dynamic loading, we first analyze the stress and displacement fields near a stationary crack tip. Because the displacements are finite, their derivatives with respect to time are also bounded for stationary cracks. In the mean time, the stresses and their derivatives with respect to spatial coordinates are expected to be singular at the crack tip. The left side terms in the equations of motion (10.50) thus dominate the inertial terms on the right side. Hence in the near-tip region, the forms of the equations of motion reduce to those of equilibrium equations as follows:

$$\frac{\partial \sigma_{xx}}{\partial x} + \frac{\partial \sigma_{xy}}{\partial y} = 0$$
$$\frac{\partial \sigma_{xy}}{\partial x} + \frac{\partial \sigma_{yy}}{\partial y} = 0 \qquad (10.60)$$

Because the material is still assumed to be linear elastic, the stress and displacement fields near the crack tip will have the same forms as those in the quasistatic linear elastic fracture mechanics introduced in Chapter 3. Hence, the inertial effect does not alter the singularity structure of stress and displacement fields near the tip of a stationary crack under dynamic loading. The crack tip stress and displacement fields are thus given by

$$\sigma_{xx} = \frac{K_I(t)}{\sqrt{2\pi r}} \cos \frac{1}{2}\theta \left(1 - \sin \frac{1}{2}\theta \sin \frac{3}{2}\theta \right)$$
$$- \frac{K_{II}(t)}{\sqrt{2\pi r}} \sin \frac{1}{2}\theta \left(2 + \cos \frac{\theta}{2} \cos \frac{3\theta}{2} \right)$$

$$\sigma_{yy} = \frac{K_I(t)}{\sqrt{2\pi r}} \cos \frac{1}{2}\theta \left(1 + \sin \frac{1}{2}\theta \sin \frac{3}{2}\theta \right)$$

$$+ \frac{K_{II}(t)}{\sqrt{2\pi r}} \sin \frac{\theta}{2} \cos \frac{\theta}{2} \cos \frac{3}{2}\theta$$

$$\sigma_{xy} = \frac{K_I(t)}{\sqrt{2\pi r}} \sin \frac{1}{2}\theta \cos \frac{1}{2}\theta \cos \frac{3}{2}\theta$$

$$+ \frac{K_{II}(t)}{\sqrt{2\pi r}} \cos \frac{1}{2}\theta \left(1 - \sin \frac{1}{2}\theta \sin \frac{3}{2}\theta \right) \tag{10.61}$$

and

$$u_x = \frac{K_I(t)}{8\mu\pi} \sqrt{2\pi r} \left[(2\kappa - 1) \cos \frac{\theta}{2} - \cos \frac{3\theta}{2} \right]$$

$$+ \frac{K_{II}(t)}{8\mu\pi} \sqrt{2\pi r} \left[(2\kappa + 3) \sin \frac{\theta}{2} + \sin \frac{3\theta}{2} \right]$$

$$u_y = \frac{K_I(t)}{8\mu\pi} \sqrt{2\pi r} \left[(2\kappa + 1) \sin \frac{\theta}{2} - \sin \frac{3\theta}{2} \right]$$

$$- \frac{K_{II}(t)}{8\mu\pi} \sqrt{2\pi r} \left[(2\kappa - 3) \cos \frac{\theta}{2} + \cos \frac{3\theta}{2} \right] \tag{10.62}$$

where $K_I(t)$ and $K_{II}(t)$ are the dynamic stress intensity factors (DSIFs) and (r,θ) are the polar coordinates centered at the crack tip with crack surfaces at $\theta = \pm\pi$. $K_I(t)$ and $K_{II}(t)$ depend not only on the magnitude of the dynamic load and crack configuration, but also time t. In general, the peak values of the DSIFs for a given crack geometry are higher than the corresponding SIFs under the quasistatic loading of the same magnitude.

The monographs by Sih [10-11] and Freund [10-12] provide the DSIF solutions for various cracks under crack face impact loads as well as wave loads. Closed-form solutions of DSIFs are available for only a few problems of dynamically loaded stationary cracks. Maue [10-13] considered a semi-infinite crack subjected to a sudden crack face pressure σ_0 at time $t = 0$ and gave the closed-form DSIF as follows:

$$K_I(t) = \frac{2\sqrt{2}}{\pi} \frac{c_R}{c_1} \sigma_0 \sqrt{\pi c_1 t} \tag{10.63}$$

where c_R is the Rayleigh surface wave speed, which is the smallest real root of the following equation:

$$\left(2 - \frac{c_R^2}{c_2^2} \right)^2 - 4 \left(1 - \frac{c_R^2}{c_1^2} \right)^{1/2} \left(1 - \frac{c_R^2}{c_2^2} \right)^{1/2} = 0 \tag{10.64}$$

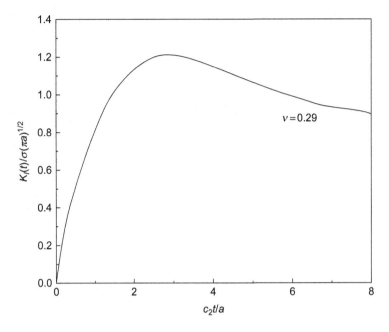

FIGURE 10.5

Mode I dynamic stress intensity factor versus nondimensional time for a central crack of length $2a$ in an infinite plate subjected to sudden crack face pressure σ (adapted from Sih et al. [10-14]).

Note that $c_R < c_2 < c_1$. Equation (10.63) shows that the DSIF depends on both dilatational and Rayleigh surface wave speeds. Moreover, the DSIF is zero at the start of loading and increases monotonically with time.

Sih et al. [10-14] investigated a central crack of length $2a$ in an infinite plate subjected to sudden crack face pressure σ at time $t = 0$. They computed the DSIF using an integral transform/integral equation approach. The numerical result is shown in Figure 10.5.

It is seen that the DSIF initially increases with time, reaches the peak value at about $c_2 t/a = 3.0$, and then decreases with time. The peak DSIF is approximately 20% percent higher than the corresponding quasistatic SIF of $\sigma \sqrt{\pi a}$. The DSIF approaches the quasistatic SIF in the steady state limit ($t \to \infty$).

In general, DSIFs have to be obtained using numerical methods, for example, the finite element method. The DSIF versus time response can be quite complicated due to wave reflections at the crack surface and structural boundaries.

The crack tip stress field Eq. (10.61) indicates that similar to the quasistatic cracks, the DSIFs determine the intensity of the singular stresses around the crack tip. Crack initiation thus can be predicted based on the SIF criterion, that is, crack initiation occurs when the DSIF reaches the dynamic fracture toughness of the material K_{Id}:

$$K_I(t) = K_{Id} \tag{10.65}$$

This fracture criterion is applicable to brittle materials such as engineering ceramics and rocks, as well as metals when small-scale yielding conditions prevail.

We know from Chapters 3 and 6 that crack tip plastic energy dissipation significantly contributes to the fracture toughness of metals. It is also known that the plastic properties of metals are rate-dependent. Metals usually become less ductile at higher loading rates, that is, the yield strength increases with increasing loading rate whereas the ductility decreases. The dynamic fracture toughness K_{Id} thus tends to decrease with increasing loading rate as the crack tip plastic deformation is supressed at higher loading rates. However, the dependence of K_{Id} on loading rate can be complex as high strain rates in the crack tip region cause temperature variation, which in turn also influences the material's constitutive behavior. K_{Id} reduces to K_{Ic} at vanishing loading rate.

10.3.3 Dynamic Crack Propagation

In a linear elastic material, a crack will propagate unstably once it has initiated. The crack propagation will lead to dynamic failure of the material unless it is arrested. Studies of the stress and displacement fields near a propagating crack tip usually employ a moving coordinate system (x_1, y_1) centered at the crack tip, as shown in Figure 10.6, where (x, y) is a fixed coordinate system and $a(t)$ is current crack propagation distance. The two systems are related by

$$x_1 = x - a(t), \quad y_1 = y \tag{10.66}$$

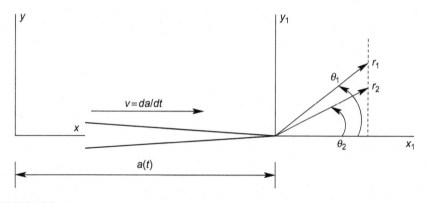

FIGURE 10.6

Coordinate systems attached to the moving crack tip.

The material derivative in the moving coordinates becomes

$$\frac{d(\ )}{dt} = \frac{\partial(\ )}{\partial t} - V\frac{\partial(\ )}{\partial x_1} \tag{10.67}$$

where $V = \dot{a} = da/dt$ is the crack propagation speed.

Using the material derivative Eq. (10.67) and coordinate transformation Eq. (10.66), the wave equations (10.58) in the moving coordinates can be written as follows:

$$\frac{\partial^2\varphi}{\partial x_1^2} + \frac{\partial^2\varphi}{\partial y_1^2} = \frac{1}{c_1^2}\left(\frac{\partial^2\varphi}{\partial t^2} - 2V\frac{\partial^2\varphi}{\partial x_1\partial t} - \dot{V}\frac{\partial\varphi}{\partial x_1} + V^2\frac{\partial^2\varphi}{\partial x_1^2}\right)$$

$$\frac{\partial^2\psi}{\partial x_1^2} + \frac{\partial^2\psi}{\partial y_1^2} = \frac{1}{c_2^2}\left(\frac{\partial^2\psi}{\partial t^2} - 2V\frac{\partial^2\psi}{\partial x_1\partial t} - \dot{V}\frac{\partial\psi}{\partial x_1} + V^2\frac{\partial^2\psi}{\partial x_1^2}\right)$$

These equations can be rewritten as

$$\alpha_1^2\frac{\partial^2\varphi}{\partial x_1^2} + \frac{\partial^2\varphi}{\partial y_1^2} = \frac{1}{c_1^2}\left(\frac{\partial^2\varphi}{\partial t^2} - 2V\frac{\partial^2\varphi}{\partial x_1\partial t} - \dot{V}\frac{\partial\varphi}{\partial x_1}\right)$$

$$\alpha_2^2\frac{\partial^2\psi}{\partial x_1^2} + \frac{\partial^2\psi}{\partial y_1^2} = \frac{1}{c_2^2}\left(\frac{\partial^2\psi}{\partial t^2} - 2V\frac{\partial^2\psi}{\partial x_1\partial t} - \dot{V}\frac{\partial\psi}{\partial x_1}\right) \tag{10.68}$$

where α_1 and α_2 are given by

$$\alpha_1 = \sqrt{1 - \left(\frac{V}{c_1}\right)^2}, \qquad \alpha_2 = \sqrt{1 - \left(\frac{V}{c_2}\right)^2} \tag{10.69}$$

Note that $\alpha_1 > \alpha_2$ as $c_1 > c_2$. When the stresses are singular at the moving crack tip, the terms on the left side of Eq. (10.68) dominate because the terms on the right side have weaker singularities. In the crack tip region, Eq. (10.68) reduce to

$$\alpha_1^2\frac{\partial^2\varphi}{\partial x_1^2} + \frac{\partial^2\varphi}{\partial y_1^2} = 0$$

$$\alpha_2^2\frac{\partial^2\psi}{\partial x_1^2} + \frac{\partial^2\psi}{\partial y_1^2} = 0 \tag{10.70}$$

The preceding equations are essentially Laplace equations in the $(x_1, \alpha_1 y_1)$ and $(x_1, \alpha_2 y_1)$ systems.

Introduce two polar coordinate systems, (r_1, θ_1) and (r_2, θ_2), at the moving crack tip as shown in Figure 10.6:

$$x_1 = r_1\cos\theta_1, \qquad y_1 = \frac{r_1}{\alpha_1}\sin\theta_1$$

$$x_1 = r_2\cos\theta_2, \qquad y_1 = \frac{r_2}{\alpha_2}\sin\theta_2 \tag{10.71}$$

or inversely:

$$r_1 = \sqrt{x_1^2 + (\alpha_1 y_1)^2}, \qquad \theta_1 = \tan^{-1}\left(\frac{\alpha_1 y_1}{x_1}\right)$$

$$r_2 = \sqrt{x_1^2 + (\alpha_2 y_1)^2}, \qquad \theta_2 = \tan^{-1}\left(\frac{\alpha_2 y_1}{x_1}\right)$$

(10.72)

It follows from these two equations that

$$\frac{\partial r_1}{\partial x_1} = \frac{x_1}{r_1} = \cos\theta_1, \qquad \frac{\partial r_1}{\partial y_1} = \frac{\alpha_1^2 y_1}{r_1} = \alpha_1 \sin\theta_1$$

$$\frac{\partial \theta_1}{\partial x_1} = -\frac{\alpha_1 y_1}{r_1^2} = -\frac{\sin\theta_1}{r_1}, \qquad \frac{\partial \theta_1}{\partial y_1} = \frac{\alpha_1 x_1}{r_1^2} = \alpha_1 \frac{\cos\theta_1}{r_1}$$

(10.73)

$$\frac{\partial r_2}{\partial x_1} = \frac{x_1}{r_2} = \cos\theta_2, \qquad \frac{\partial r_2}{\partial y_1} = \frac{\alpha_2^2 y_1}{r_2} = \alpha_2 \sin\theta_2$$

$$\frac{\partial \theta_2}{\partial x_1} = -\frac{\alpha_2 y_1}{r_2^2} = -\frac{\sin\theta_2}{r_2}, \qquad \frac{\partial \theta_2}{\partial y_1} = \frac{\alpha_2 x_1}{r_2^2} = \alpha_2 \frac{\cos\theta_2}{r_2}$$

(10.74)

Thus,

$$\frac{\partial(\ )}{\partial x_1} = \frac{\partial(\ )}{\partial r_1}\cos\theta_1 - \frac{\partial(\ )}{\partial \theta_1}\frac{\sin\theta_1}{r_1}$$

$$\frac{\partial(\ )}{\partial y_1} = \frac{\partial(\ )}{\partial r_1}\alpha_1 \sin\theta_1 + \frac{\partial(\ )}{\partial \theta_1}\alpha_1\frac{\cos\theta_1}{r_1}$$

(10.75)

in the (r_1,θ_1) system, and

$$\frac{\partial(\ )}{\partial x_1} = \frac{\partial(\ )}{\partial r_2}\cos\theta_2 - \frac{\partial(\ )}{\partial \theta_2}\frac{\sin\theta_2}{r_2}$$

$$\frac{\partial(\ )}{\partial y_1} = \frac{\partial(\ )}{\partial r_2}\alpha_2 \sin\theta_2 + \frac{\partial(\ )}{\partial \theta_2}\alpha_2\frac{\cos\theta_2}{r_2}$$

(10.76)

in the (r_2,θ_2) system. Since $\varphi = \varphi(r_1,\theta_1)$ and $\psi = \psi(r_2,\theta_2)$, Eq. (10.70) can be written in the moving polar coordinates as follows:

$$\frac{\partial^2 \varphi}{\partial r_1^2} + \frac{1}{r_1}\frac{\partial \varphi}{\partial r_1} + \frac{1}{r_1^2}\frac{\partial^2 \varphi}{\partial \theta_1^2} = 0$$

$$\frac{\partial^2 \psi}{\partial r_2^2} + \frac{1}{r_2}\frac{\partial \psi}{\partial r_2} + \frac{1}{r_2^2}\frac{\partial^2 \psi}{\partial \theta_2^2} = 0$$

(10.77)

These are Laplace equations in the polar (r_1,θ_1) and (r_2,θ_2) coordinate systems, respectively.

To obtain the asymptotic singular stresses at the moving crack tip, the displacement potential functions are assumed to have the following form:

$$\varphi = r_1^s \widetilde{\varphi}(\theta_1), \qquad r_1 \to 0$$
$$\psi = r_2^s \widetilde{\psi}(\theta_2), \qquad r_2 \to 0 \tag{10.78}$$

where s is the eigenvalue to be determined by the boundary conditions. Substituting these potentials into Eq. (10.77), we have

$$\frac{d^2\widetilde{\varphi}}{d\theta_1^2} + s^2\widetilde{\varphi} = 0$$

$$\frac{d^2\widetilde{\psi}}{d\theta_2^2} + s^2\widetilde{\psi} = 0$$

The general solutions of these equations are

$$\widetilde{\varphi} = C_1\cos(s\theta_1) + C_3\sin(s\theta_1)$$
$$\widetilde{\psi} = C_4\cos(s\theta_2) + C_2\sin(s\theta_2)$$

where C_i ($i = 1,2,3,4$) are constants. In the following we consider Mode I crack propagation only. Mode I symmetry consideration gives $C_3 = C_4 = 0$. Hence, $\widetilde{\varphi}$ and $\widetilde{\psi}$ reduce to

$$\widetilde{\varphi} = C_1\cos(s\theta_1)$$
$$\widetilde{\psi} = C_2\sin(s\theta_2)$$

Substituting these expressions into Eq. (10.78), and using the relations Eqs. (10.57), (10.55), and (10.73) through (10.76), we obtain the following dominant asymptotic expressions for the displacement potentials, displacements, and stresses at the moving crack tip ($r_1 \to 0$, $r_2 \to 0$):

$$\varphi = C_1 r_1^s \cos(s\theta_1)$$
$$\psi = C_2 r_2^s \sin(s\theta_2) \tag{10.79}$$

$$u_x = sC_1 r_1^{s-1}\cos(s-1)\theta_1 + \alpha_2 s C_2 r_2^{s-1}\cos(s-1)\theta_2$$
$$u_y = -\alpha_1 s C_1 r_1^{s-1}\sin(s-1)\theta_1 - s C_2 r_2^{s-1}\sin(s-1)\theta_2 \tag{10.80}$$

and

$$\sigma_{xx} = \rho c_2^2 (s^2 - s)(1 + 2\alpha_1^2 - \alpha_2^2)C_1 r_1^{s-2}\cos(s-2)\theta_1$$
$$\qquad + 2\rho c_2^2(s^2 - s)\alpha_2 C_2 r_2^{s-2}\cos(s-2)\theta_2$$
$$\sigma_{yy} = -\rho c_2^2(s^2 - s)(1 + \alpha_2^2)C_1 r_1^{s-2}\cos(s-2)\theta_1$$

$$-2\rho c_2^2(s^2-s)\alpha_2 C_2 r_2^{s-2}\cos(s-2)\theta_2$$

$$\sigma_{xy} = -2\rho c_2^2(s^2-s)\alpha_1 C_1 r_1^{s-2}\sin(s-2)\theta_1$$

$$-\rho c_2^2(s^2-s)(1+\alpha_2^2)C_2 r_2^{s-2}\sin(s-2)\theta_2 \tag{10.81}$$

The preceding stresses and displacements already satisfy the symmetry conditions along the crack line, that is,

$$\sigma_{xy} = u_y = 0, \qquad \theta_1 = \theta_2 = 0$$

The crack surface boundary conditions for the analysis of crack tip asymptotic solutions are

$$\sigma_{xy} = \sigma_{yy} = 0, \qquad \theta_1 = \theta_2 = \pi \tag{10.82}$$

Substituting the stress expressions in Eq. (10.81) into the boundary conditions in Eq. (10.82) yields two equations satisfied by constants C_1 and C_2 as follows ($r_1 = r_2$ at $\theta_1 = \theta_2 = \pi$):

$$-2\rho c_2^2(s^2-s)\alpha_1 C_1 \sin(s-2)\pi - \rho c_2^2(s^2-s)(1+\alpha_2^2)C_2 \sin(s-2)\pi = 0$$

$$-\rho c_2^2(s^2-s)(1+\alpha_2^2)C_1 \cos(s-2)\pi - 2\rho c_2^2(s^2-s)\alpha_2 C_2 \cos(s-2)\pi = 0 \tag{10.83}$$

For the nontrivial solutions to exist, the determinant of the equation systems here must be zero, that is,

$$\rho c_2^2(s^2-s)\left[4\alpha_1\alpha_2 - (1+\alpha_2^2)^2\right]\cos[(s-2)\pi]\sin[(s-2)\pi]$$

$$= \frac{1}{2}\rho c_2^2(s^2-s)\left[4\alpha_1\alpha_2 - (1+\alpha_2^2)^2\right]\sin[2(s-2)\pi] = 0 \tag{10.84}$$

Equation (10.84) is the characteristic equation for determination of the eigenvalue s. Its root that leads to both finite strain energy and stress singularity is

$$s = 3/2$$

There is only one independent equation in Eq. (10.83), which gives the relation between constants C_1 and C_2 as follows:

$$C_2 = -\frac{2\alpha_1}{1+\alpha_2^2}C_1$$

The constants C_1 can be related to the Mode I dynamic SIF $K_I(t)$:

$$C_1 = \frac{4(1+\alpha_2^2)}{\mu\left[4\alpha_1\alpha_2 - (1+\alpha_2^2)^2\right]}\frac{K_I(t)}{3\sqrt{2\pi}}$$

with $K_I(t)$ defined by

$$K_I(t) = \lim_{r_1\to 0} \sqrt{2\pi r_1}\sigma_{yy}(r_1,r_2,\theta_1,\theta_2)|_{\theta_1=\theta_2=0,r_1=r_2}$$

The crack tip dominant stress and displacement fields can thus be written in the following form:

$$
\sigma_{xx} = \frac{K_I(t)}{\sqrt{2\pi}} \frac{1+\alpha_2^2}{4\alpha_1\alpha_2 - (1+\alpha_2^2)^2} \left[\frac{1+2\alpha_1^2-\alpha_2^2}{\sqrt{r_1}} \cos\frac{\theta_1}{2} - \frac{4\alpha_1\alpha_2}{(1+\alpha_2^2)\sqrt{r_2}} \cos\frac{\theta_2}{2} \right]
$$

$$
\sigma_{yy} = \frac{K_I(t)}{\sqrt{2\pi}} \frac{1+\alpha_2^2}{4\alpha_1\alpha_2 - (1+\alpha_2^2)^2} \left[-\frac{1+\alpha_2^2}{\sqrt{r_1}} \cos\frac{\theta_1}{2} + \frac{4\alpha_1\alpha_2}{(1+\alpha_2^2)\sqrt{r_2}} \cos\frac{\theta_2}{2} \right]
$$

$$
\sigma_{xy} = \frac{K_I(t)}{\sqrt{2\pi}} \frac{2\alpha_1(1+\alpha_2^2)}{4\alpha_1\alpha_2 - (1+\alpha_2^2)^2} \left[\frac{1}{\sqrt{r_1}} \sin\frac{\theta_1}{2} - \frac{1}{\sqrt{r_2}} \sin\frac{\theta_2}{2} \right] \tag{10.85}
$$

$$
u_x = \frac{K_I(t)}{\sqrt{2\pi}} \frac{2(1+\alpha_2^2)}{\mu\left[4\alpha_1\alpha_2 - (1+\alpha_2^2)^2\right]} \left[\sqrt{r_1}\cos\frac{\theta_1}{2} - \frac{2\alpha_1\alpha_2}{(1+\alpha_2^2)}\sqrt{r_2}\cos\frac{\theta_2}{2} \right]
$$

$$
u_y = \frac{K_I(t)}{\sqrt{2\pi}} \frac{2(1+\alpha_2^2)}{\mu\left[4\alpha_1\alpha_2 - (1+\alpha_2^2)^2\right]} \left[-\alpha_1\sqrt{r_1}\sin\frac{\theta_1}{2} + \frac{2\alpha_1}{(1+\alpha_2^2)}\sqrt{r_2}\sin\frac{\theta_2}{2} \right] \tag{10.86}
$$

The crack tip dominant velocity field can also be obtained by using

$$
\frac{d(\ )}{dt} = -V\frac{\partial(\ )}{\partial x_1}, \qquad r_1, r_2 \to 0,
$$

and the result is

$$
\dot{u}_x = -\frac{VK_I(t)}{\sqrt{2\pi}} \frac{(1+\alpha_2^2)}{\mu\left[4\alpha_1\alpha_2 - (1+\alpha_2^2)^2\right]} \left[\frac{1}{\sqrt{r_1}}\cos\frac{\theta_1}{2} - \frac{2\alpha_1\alpha_2}{(1+\alpha_2^2)}\frac{1}{\sqrt{r_2}}\cos\frac{\theta_2}{2} \right]
$$

$$
\dot{u}_y = -\frac{VK_I(t)}{\sqrt{2\pi}} \frac{(1+\alpha_2^2)}{\mu\left[4\alpha_1\alpha_2 - (1+\alpha_2^2)^2\right]} \left[\frac{\alpha_1}{\sqrt{r_1}}\sin\frac{\theta_1}{2} - \frac{2\alpha_1}{(1+\alpha_2^2)}\frac{1}{\sqrt{r_2}}\sin\frac{\theta_2}{2} \right] \tag{10.87}
$$

It can be seen that the velocity field has an inverse square root singularity at the moving crack tip. The stress and displacement fields in Eq. (10.87) under constant crack propagation speed were derived by Rice [10-15]. Freund and Clifton [10-16] and Nielson [10-17] later showed that the results are also valid when the propagation speed varies with time.

Equations (10.85) and (10.86) show that the stresses at a propagating crack tip still have the inverse square singularity. The angular distributions of the stresses and displacements, however, depend on the crack propagation speed. It is interesting to look at the stress distributions along the crack line ($\theta_1 = \theta_2 = 0$, $r_1 = r_2 = x_1 > 0$). It follows from Eq. (10.85) that

$$\sigma_{xx}|_{\theta_1=\theta_2=0} = \frac{K_I(t)}{\sqrt{2\pi x_1}} \frac{\left(1+\alpha_2^2\right)\left(1+2\alpha_1^2-\alpha_2^2\right) - 4\alpha_1\alpha_2}{4\alpha_1\alpha_2 - \left(1+\alpha_2^2\right)^2}$$

$$\sigma_{yy}|_{\theta_1=\theta_2=0} = \frac{K_I(t)}{\sqrt{2\pi x_1}}$$

The ratio of σ_{yy} to σ_{xx} at the crack line is

$$\left(\frac{\sigma_{yy}}{\sigma_{xx}}\right)_{\theta_1=\theta_2=0} = \frac{4\alpha_1\alpha_2 - \left(1+\alpha_2^2\right)^2}{\left(1+\alpha_2^2\right)\left(1+2\alpha_1^2-\alpha_2^2\right) - 4\alpha_1\alpha_2} \qquad (10.88)$$

This ratio represents the stress triaxiality ahead of the propagating crack. This ratio is unity for stationary cracks. For rapidly propagating cracks, the ratio varies with the crack speed.

Figure 10.7 schematically shows the variation of the stress ratio versus the nondimensional crack propagation speed V/c_R. It is seen that the stress ratio approaches unity when the crack propagation speed goes to zero, consistent with the stationary crack result. The ratio decreases monotonically with increasing crack speed and approaches zero when the crack speed reaches the Rayleigh surface wave speed. In fracture mechanics, we know that the fracture toughness increases with decreasing triaxiality of the stress field. Equation (10.88) and Figure 10.7 imply that dynamic fracture toughness will increase with increasing crack propagation speed due to the stress ratio effect.

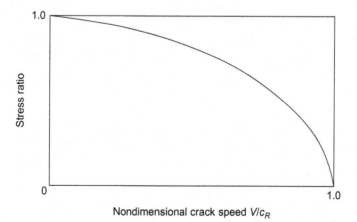

FIGURE 10.7

Stress ratio versus crack propagation speed ahead of the crack.

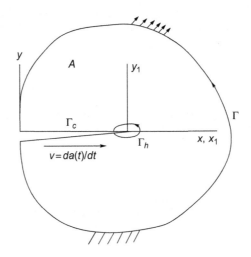

FIGURE 10.8

Dynamic crack propagation and the contours around the moving crack tip.

It is known from Chapter 4 that energy release rate is an important concept in fracture mechanics. To introduce the dynamic energy release rate for a rapidly propagating crack, we consider the flux of energy, F, into the crack tip region through a vanishingly small contour Γ_h around the moving crack tip as shown in Figure 10.8, where Γ is the outer boundary of the cracked body and A is the area bounded by Γ, Γ_h, and the crack faces Γ_c between Γ and Γ_h. The medium is subjected to the prescribed traction T_i along the boundary segment Γ_t and the prescribed displacements on the boundary segment Γ_u. Moreover, the small contour Γ_h is fixed relative to, and moves with the crack tip. Clearly the energy flux F equals the rate of work done by the traction on Γ_t less the rates of increases in the strain energy and kinetic energy, that is,

$$F = \int_{\Gamma_t} T_i \dot{u}_i d\Gamma - \lim_{\Gamma_h \to 0} \frac{d}{dt} \iint_A W dA - \lim_{\Gamma_h \to 0} \frac{d}{dt} \iint_A \frac{1}{2} \rho \dot{u}_i \dot{u}_i dA \qquad (10.89)$$

Using the transport theorem, we have

$$\frac{d}{dt} \iint_A W dA = \iint_A \dot{W} dA - \int_{\Gamma_h} W V n_1 d\Gamma$$

$$\frac{d}{dt} \iint_A \frac{1}{2} \rho \dot{u}_i \dot{u}_i dA = \iint_A \rho \ddot{u}_i \dot{u}_i dA - \int_{\Gamma_h} \frac{1}{2} \rho \dot{u}_i \dot{u}_i V n_1 d\Gamma$$

where n_1 is the x-component of the outward unit normal to Γ_h. Substituting these equations in (10.89), we have

$$F = \int_{\Gamma} T_i \dot{u}_i d\Gamma + \lim_{\Gamma_h \to 0} \int_{\Gamma_h} \left(W + \frac{1}{2}\rho \dot{u}_i \dot{u}_i \right) V n_1 d\Gamma - \lim_{\Gamma_h \to 0} \iint_{A} (\dot{W} + \rho \ddot{u}_i \dot{u}_i) dA$$

(10.90)

where the integral along Γ_t has been extended to Γ because $\dot{u}_i$ is zero along Γ_u. Using the definition of W and the equations of motion, we know

$$\dot{W} = \sigma_{ij} \dot{e}_{ij} = \sigma_{ij} \dot{u}_{i,j}$$

$$\sigma_{ij} \dot{u}_{i,j} = (\sigma_{ij} \dot{u}_i)_{,j} - \sigma_{ij,j} \dot{u}_i = (\sigma_{ij} \dot{u}_i)_{,j} - \rho \ddot{u}_i \dot{u}_i$$

Use of these relations in the area integral on the right side of Eq. (10.90) and application of the divergence theorem yield

$$\iint_{A} (\dot{W} + \rho \ddot{u}_i \dot{u}_i) dA = \iint_{A} (\sigma_{ij} \dot{u}_i)_{,j} dA = \int_{\Gamma + \Gamma_c - \Gamma_h} T_i \dot{u}_i d\Gamma = \int_{\Gamma - \Gamma_h} T_i \dot{u}_i d\Gamma$$

Substituting this into Eq. (10.90), we obtain the energy flux as follows:

$$F = \lim_{\Gamma_h \to 0} \int_{\Gamma_h} \left[\left(W + \frac{1}{2}\rho \dot{u}_i \dot{u}_i \right) V n_1 + T_i \dot{u}_i \right] d\Gamma$$

(10.91)

The energy release rate is clearly F/V as $V = da/dt$, and is given by

$$G = \frac{F}{V} = \lim_{\Gamma_h \to 0} \frac{1}{V} \int_{\Gamma_h} \left[\left(W + \frac{1}{2}\rho \dot{u}_i \dot{u}_i \right) V n_1 + T_i \dot{u}_i \right] d\Gamma$$

(10.92)

Under steady state conditions, V is a constant and the material derivative in Eq. (10.67) reduces to

$$\frac{d(\ )}{dt} = -V \frac{\partial (\ )}{\partial x_1}$$

The energy release rate Eq. (10.92) becomes

$$G = \lim_{\Gamma_h \to 0} \int_{\Gamma_h} \left[\left(W + \frac{1}{2}\rho V^2 \frac{\partial u_i}{\partial x_1} \frac{\partial u_i}{\partial x_1} \right) n_1 - T_i \frac{\partial u_i}{\partial x_1} \right] d\Gamma$$

(10.93)

It can be shown (Atkinson and Eshelby [10-25]) that this integral is a path-independent integral for any contour around the crack tip, beginning at the lower crack surface and ending on the upper crack surface.

Similar to quasistatic cracks, there is a relationship between the DSIF and the dynamic energy release rate G. For plane strain, the relation was obtained by Freund [10-18] and Nilsson [10-19] as follows:

$$G_I(t) = \frac{1 - \nu^2}{E} A(V) K_I^2(t) \tag{10.94}$$

where the dynamic factor $A(V)$ is given by

$$A(V) = \frac{1}{1 - \nu} \frac{\alpha_1 \left(1 - \alpha_2^2\right)}{4\alpha_1 \alpha_2 - (1 + \alpha_2^2)^2}$$

Figure 10.9 schmeatically shows the variation of factor $A(V)$ with nondimensional crack speed V/c_R. It is seen that $A(V)$ goes to unity in the limit of zero crack speed. The dynamic relation Eq. (10.94) thus reduces to that in the quasistatic fracture mechanics. $A(V)$ increases monotonically with crack speed and goes to infinity when the crack speed approaches the Rayleigh wave speed implying that the maximum crack propagation speed for Mode I cracks is the Rayleigh surface wave speed.

Equation (10.85) indicates that the DSIF is the governing parameter of the crack tip singular stress field. Hence, continuous rapid crack propagation occurs when the DSIF equals its critical value, or fracture propagation toughness, K_{ID}:

$$K_I(t) = K_{ID}(V) \tag{10.95}$$

K_{ID} is a material property dependent on crack speed. Note that K_{ID} is different from K_{Id}, the dynamic fracture toughness for crack initiation in Eq. (10.65). Moreover, $K_{ID}(0) = \lim_{V \to 0} K_{ID}(V)$ is generally not equal to K_{Ic}.

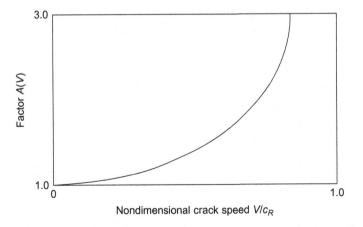

FIGURE 10.9

Variation of factor $A(V)$ with nondimensional crack propagation speed.

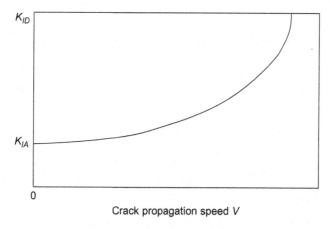

FIGURE 10.10

Schematic of relationship between fracture propagation toughness K_{ID} and crack propagation speed V.

Experimental investigations of Kanazawa and Machida [10-20] and Rosakis and Freund [10-21] for steels showed that K_{ID} is relatively insensitive to V at low crack speeds and increases dramatically with an increase in crack speed when the speed approaches a limiting, material-dependent value. This kind of behavior of K_{ID} as a function of crack speed is schematically shown in Figure 10.10.

Figure 10.10 shows that the fracture propagation toughness K_{ID} approaches K_{IA} in the limit of vanishing crack speed. K_{IA} is generally called fracture arrest toughness and is the minimum value of $K_{ID}(V)$. The crack arrest occurs when the DSIF drops below K_{IA}, that is,

$$K_I(t) < K_{IA} \tag{10.96}$$

10.3.4 Yoffe Crack

As a result of mathematical complexities, closed-form solutions are available for only a few rapid crack propagation problems. Yoffe [10-22] considered a simplified crack model in which a crack of constant length $2a$ propagates at a constant speed V in an infinite plate subjected to a remote tensile loading σ as shown in Figure 10.11. Clearly Yoffe's model is not realistic as the left tip of the crack is required to close during propagation to maintain a constant crack length, which implies that the SIF should be equal to zero at the left tip.

Yoffe obtained the solution of the problem using a Fourier transform method. Here we introduce a complex potential technique for the solution. We begin with Eq. (10.68) for the displacement potentials, which are still valid but now we assume that the rectangular coordinate system (x_1, y_1) is attached to the center of the moving crack with the crack tips at $x_1 = \pm a$, $y_1 = 0$, respectively. Using the superposition method, the boundary conditions for the Yoffe problem in the moving coordinate

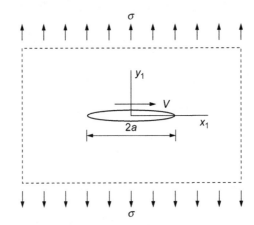

FIGURE 10.11

Yoffe crack: a crack of constant length $2a$ propagating at a constant speed V in an infinite plate.

system can be formulated as follows:

$$\sigma_{yy} = -\sigma, \quad \sigma_{xy} = 0, \qquad 0 \le |x_1| \le a, \; y_1 = 0$$
$$u_y = 0, \quad \sigma_{xy} = 0, \qquad a < |x_1| < \infty, \; y_1 = 0$$
$$\sigma_{xx}, \, \sigma_{yy}, \, \sigma_{xy} \to 0, \qquad \sqrt{x_1^2 + y_1^2} \to \infty \tag{10.97}$$

Under steady state crack propagation conditions, the terms on the right side of Eq. (10.68) vanish and the equations reduce to

$$\frac{\partial^2 \varphi}{\partial x_1^2} + \frac{\partial^2 \varphi}{\partial(\alpha_1 y_1)^2} = 0$$
$$\frac{\partial^2 \psi}{\partial x_1^2} + \frac{\partial^2 \psi}{\partial(\alpha_2 y_1)^2} = 0 \tag{10.98}$$

These are Laplace equations in the coordinate systems of $(x_1, \alpha_1 y_1)$ and $(x_1, \alpha_2 y_1)$, respectively. Hence, φ can be associated with the real part of a complex function $F_1(z_1)$, and ψ with the imaginary part of a complex function $F_2(z_2)$, that is,

$$\varphi = \varphi(x_1, \alpha_1 y_1) = F_1(z_1) + \overline{F_1(z_1)}$$
$$\psi = \psi(x_1, \alpha_2 y_1) = i\left[F_2(z_2) - \overline{F_2(z_2)} \right] \tag{10.99}$$

where a bar over a quantity denotes its complex conjugate, and

$$z_1 = x_1 + i(\alpha_1 y_1)$$
$$z_2 = x_1 + i(\alpha_2 y_1)$$

(10.100)

Using these complex potentials, the displacements and stresses can be expressed in the following forms (Gladwell [10-23]):

$$u_x + iu_y = (1 - \alpha_1)\Phi(z_1) + (1 + \alpha_1)\overline{\Phi(z_1)}$$
$$+ (1 - \alpha_2)\Psi(z_2) - (1 + \alpha_2)\overline{\Psi(z_2)}$$

(10.101)

$$\sigma_{xx} + \sigma_{yy} = 2\mu\left(\alpha_1^2 - \alpha_2^2\right)\left[\Phi'(z_1) + \overline{\Phi'(z_1)}\right]$$
$$\sigma_{xx} - \sigma_{yy} + 2i\sigma_{xy} = 2\mu\left[(1 - \alpha_1)^2\Phi'(z_1) + (1 + \alpha_1)^2\overline{\Phi'(z_1)}\right.$$
$$\left. + (1 - \alpha_2)^2\Psi'(z_2) - (1 + \alpha_2)^2\overline{\Psi'(z_2)}\right]$$

(10.102)

where

$$\Phi(z_1) = F_1'(z_1), \quad \Psi(z_2) = F_2'(z_2)$$

For the Yoffe crack problem, the complex potential functions $\Phi(z_1)$ and $\Psi(z_2)$ can be chosen as (Fan [10-24])

$$\Phi'(z_1) = \frac{\sigma}{2\mu}\frac{1 + \alpha_2^2}{4\alpha_1\alpha_2 - \left(1 + \alpha_2^2\right)^2}\left(\frac{z_1}{\sqrt{z_1^2 - a^2}} - 1\right)$$

$$\Psi'(z_2) = \frac{\sigma}{2\mu}\frac{2\alpha_1}{4\alpha_1\alpha_2 - \left(1 + \alpha_2^2\right)^2}\left(\frac{z_2}{\sqrt{z_2^2 - a^2}} - 1\right)$$

(10.103)

Substituting this into the stress expressions Eq. (10.102) along the crack line ($x_1 > a$, $y_1 = 0$, $z_1 = z_2 = x_1$), we have

$$\sigma_{xx} + \sigma_{yy} = 2\mu\left(\alpha_1^2 - \alpha_2^2\right)\frac{2\sigma}{2\mu}\frac{1 + \alpha_2^2}{4\alpha_1\alpha_2 - \left(1 + \alpha_2^2\right)^2}\left(\frac{x_1}{\sqrt{x_1^2 - a^2}} - 1\right)$$

$$\sigma_{xx} - \sigma_{yy} + 2i\sigma_{xy} = 2\mu\left[2 + 2\alpha_1^2\right]\frac{\sigma}{2\mu}\frac{1 + \alpha_2^2}{4\alpha_1\alpha_2 - \left(1 + \alpha_2^2\right)^2}\left(\frac{x_1}{\sqrt{x_1^2 - a^2}} - 1\right)$$

$$2\mu\left[-4\alpha_2\right]\frac{\sigma}{2\mu}\frac{2\alpha_1}{4\alpha_1\alpha_2 - \left(1 + \alpha_2^2\right)^2}\left(\frac{x_1}{\sqrt{x_1^2 - a^2}} - 1\right)$$

(10.104)

The normal stress σ_{yy} along the crack line can be obtained as follows:

$$\sigma_{yy}(x_1, 0) = \sigma \left[\frac{x_1}{\sqrt{x_1^2 - a^2}} - 1 \right], \quad x_1 > a$$

which is independent of crack propagation speed. Hence, the DSIFs at both crack tips are also independent of crack speed and have the same expression as that for the corresponding quasistatic crack, that is,

$$K_I = \lim_{x_1 \to a^+} \sqrt{2\pi(x_1 - a)}\sigma_{yy}(x_1, 0) = \sigma\sqrt{\pi a} \qquad (10.105)$$

References

[10-1] S.G. Lekhnitskii (translated by P. Fern), Theory of Elasticity of an Anisotropic Elastic Body, Holden-Day, San Francisco, 1963.

[10-2] G.C. Sih, H. Liebowitz, Mathematical theories of brittle fracture, in: H. Liebowitz (Ed.), Fracture, Vol. 2, Academic Press, New York, 1968, pp. 67–190.

[10-3] G.J. Weng, Some elastic properties of reinforced solids with special reference to isotropic ones containing spherical inclusions, Int. J. Eng. Sci. 22 (1984) 845–856.

[10-4] Z.-H. Jin, N. Noda, Crack-tip singular fields in nonhomogeneous materials, J. Appl. Mech. 61 (1994) 738–740.

[10-5] M.L. Williams, On the stress distribuion at the base of a stationary crack, J. Appl. Mech. 24 (1957) 109–114.

[10-6] J.W. Eischen, Fracture of nonhomogeneous materials, Int. J. Fract. 34 (1987) 3–22.

[10-7] Z.-H. Jin, R.C. Batra, Some basic fracture mechanics concepts in functionally graded materials, J. Mech. Phys. Sol. 44 (1996) 1221–1235.

[10-8] Z.-H. Jin, C.T. Sun, Integral representation of energy release rate in graded materials, J. Appl. Mech. 74 (2007) 1046–1048.

[10-9] T. Honein, G. Herrmann, Conservation laws in nonhomogeneous plane elastostatics, J. Mech. Phys. Sol. 45 (1997) 789–805.

[10-10] F. Delale, F. Erdogan, On the mechanical modeling of an interfacial region in bonded half-planes, J. Appl. Mech. 55 (1988) 317–324.

[10-11] G.C. Sih, Mechanics of Fracture, Vol. 4: Elastodynamic Crack Problems, Noordhoff International Publishing, Leyden, 1977.

[10-12] L.B. Freund, Dynamic Fracture Mechanics, Cambridge University Press, Cambridge, UK, 1990.

[10-13] A.W. Maue, Die entspannungswelle bei plotzlichem Einschnitt eines gespannten elastischen Korpers, Zeitschrift fur angewandte Mathematik und Mechanik, 34 (1954) 1–12.

[10-14] G.C. Sih, G.T. Embley, R.S. Ravera, Impact response of a finite crack in plane extension, Int. J. Sol. Struct. 8 (1972) 977–993.

[10-15] J.R. Rice, Mathematical analysis in the mechanics of fracture, in: H. Liebowitz (Ed.), Fracture, Vol. 2, Academic Press, New York, 1968, pp. 191–311.

[10-16] L.B. Freund, R.J. Clifton, On the uniqueness of plane elastodynamic solutions for running cracks, J. Elast. 4 (1974) 293–299.

[10-17] F. Nilsson, A note on the stress singularity at a non-uniformly moving crack tip, J. Elast. 4 (1974) 73–75.

[10-18] L.B. Freund, Crack propagation in an elastic solid subjected to general loading 2, nonuniform rate of extension, J. Mech. Phys. Sol. 20 (1972) 141–152.

[10-19] F. Nilsson, Dynamic stress intensity factors for finite strip problems, Int. J. Fract. Mech. 8 (1972) 403–411.

[10-20] T. Kanazawa, S. Machida, Fracture dynamics analysis on fast fracture and crack arrest experiments, in: T. Kanazawa, A.S. Kobayashi, K. Ido (Eds.), Fracture Tolerance Evaluation, Toyoprint, Japan, 1982.

[10-21] A.J. Rosakis, L.B. Freund, Optical measurement of the plastic strain concentration at a crack tip in a ductile steel plate, J. Eng. Mater. Technol. 104 (1982) 115–120.

[10-22] E.H. Yoffe, The moving Griffith crack, Philos. Mag. 42 (1951) 739–750.

[10-23] G.M.L. Gladwell, On the solution of problems of dynamic plane elasticity, Mathematika 4 (1957) 166–168.

[10-24] T.Y. Fan, Moving Dugdale model, Zeitschrift fur angewante Mathematik und Physik 38 (1987) 630–641.

[10-25] C. Atkinson, J.D. Eshelby, The flow of energy into the tip of a moving crack, Int. J. Fract. Mech. 4 (1968) 3–8.

Stress Intensity Factors

This appendix lists the stress intensity factors for some typical cracked specimens and structures from the handbook by Tada et al. (H. Tada, P. C. Paris, and G. R. Irwin, *The Stress Analysis of Cracks Handbook*, New York, ASME Press, 2000) which provides comprehensive stress intensity factor solutions for various crack geometries and loadings.

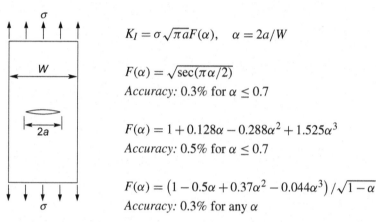

$$K_I = \sigma \sqrt{\pi a} F(\alpha), \quad \alpha = 2a/W$$

$$F(\alpha) = \sqrt{\sec(\pi\alpha/2)}$$
Accuracy: 0.3% for $\alpha \le 0.7$

$$F(\alpha) = 1 + 0.128\alpha - 0.288\alpha^2 + 1.525\alpha^3$$
Accuracy: 0.5% for $\alpha \le 0.7$

$$F(\alpha) = \left(1 - 0.5\alpha + 0.37\alpha^2 - 0.044\alpha^3\right)/\sqrt{1-\alpha}$$
Accuracy: 0.3% for any α

Center-Cracked Tension Plate

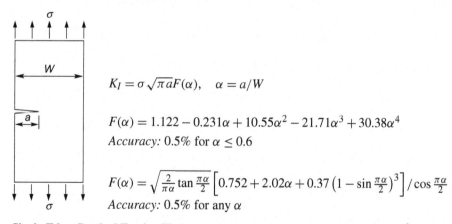

$$K_I = \sigma \sqrt{\pi a} F(\alpha), \quad \alpha = a/W$$

$$F(\alpha) = 1.122 - 0.231\alpha + 10.55\alpha^2 - 21.71\alpha^3 + 30.38\alpha^4$$
Accuracy: 0.5% for $\alpha \le 0.6$

$$F(\alpha) = \sqrt{\frac{2}{\pi\alpha} \tan \frac{\pi\alpha}{2}} \left[0.752 + 2.02\alpha + 0.37 \left(1 - \sin \frac{\pi\alpha}{2}\right)^3\right]/\cos \frac{\pi\alpha}{2}$$
Accuracy: 0.5% for any α

Single-Edge-Cracked Tension Plate

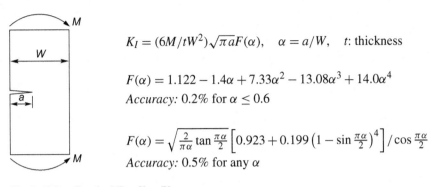

$$K_I = (6M/tW^2)\sqrt{\pi a}F(\alpha), \quad \alpha = a/W, \quad t: \text{thickness}$$

$$F(\alpha) = 1.122 - 1.4\alpha + 7.33\alpha^2 - 13.08\alpha^3 + 14.0\alpha^4$$
Accuracy: 0.2% for $\alpha \leq 0.6$

$$F(\alpha) = \sqrt{\frac{2}{\pi\alpha} \tan\frac{\pi\alpha}{2}} \left[0.923 + 0.199 \left(1 - \sin\frac{\pi\alpha}{2} \right)^4 \right] / \cos\frac{\pi\alpha}{2}$$
Accuracy: 0.5% for any α

Single-Edge-Cracked Bending Plate

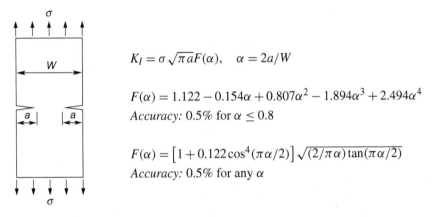

$$K_I = \sigma\sqrt{\pi a}F(\alpha), \quad \alpha = 2a/W$$

$$F(\alpha) = 1.122 - 0.154\alpha + 0.807\alpha^2 - 1.894\alpha^3 + 2.494\alpha^4$$
Accuracy: 0.5% for $\alpha \leq 0.8$

$$F(\alpha) = \left[1 + 0.122\cos^4(\pi\alpha/2) \right] \sqrt{(2/\pi\alpha)\tan(\pi\alpha/2)}$$
Accuracy: 0.5% for any α

Double-Edge-Cracked Tension Plate

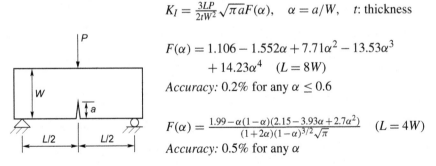

$$K_I = \frac{3LP}{2tW^2}\sqrt{\pi a}F(\alpha), \quad \alpha = a/W, \quad t: \text{thickness}$$

$$F(\alpha) = 1.106 - 1.552\alpha + 7.71\alpha^2 - 13.53\alpha^3$$
$$+ 14.23\alpha^4 \quad (L = 8W)$$
Accuracy: 0.2% for any $\alpha \leq 0.6$

$$F(\alpha) = \frac{1.99 - \alpha(1-\alpha)(2.15 - 3.93\alpha + 2.7\alpha^2)}{(1+2\alpha)(1-\alpha)^{3/2}\sqrt{\pi}} \quad (L = 4W)$$
Accuracy: 0.5% for any α

Three-Point Bending Specimen

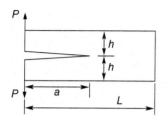

$$K_I = (P/t\sqrt{h})F(\alpha), \quad \alpha = a/L, \quad t: \text{thickness}$$

$$F(\alpha) = 2\sqrt{3}(a/h + 0.64)$$
Accuracy: 1% for $2 \leq a/h \leq 10$

Double Cantilever Beam

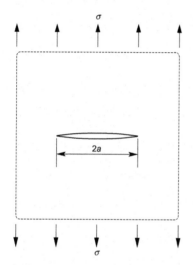

$$K_I = \sigma\sqrt{\pi a}$$

Single Crack in an Infinite Plate Subjected to Tension

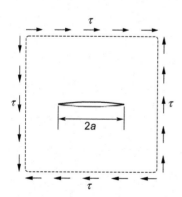

$$K_{II} = \tau\sqrt{\pi a}$$

Single Crack in an Infinite Plate Subjected to Shear

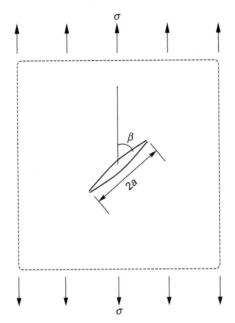

$$K_I = \sigma \sqrt{\pi a} \sin^2 \beta$$
$$K_{II} = \sigma \sqrt{\pi a} \sin \beta \cos \beta$$

Single Inclined Crack in an Infinite Plate Subjected to Tension

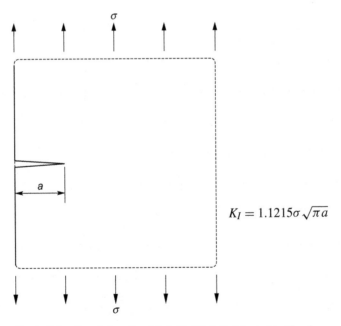

$$K_I = 1.1215 \sigma \sqrt{\pi a}$$

Single Edge Crack in a Semi-infinite Plate Subjected to Tension

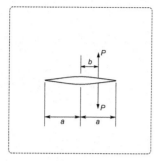

$$K_I = \frac{P}{\sqrt{\pi a}}\sqrt{\frac{a+b}{a-b}} \quad \text{(right crack tip)}$$

$$K_I = \frac{P}{\sqrt{\pi a}}\sqrt{\frac{a-b}{a+b}} \quad \text{(left crack tip)}$$

Single Crack in an Infinite Plate Subjected to a Pair of Concentrated Compressive Forces (per unit thickness) on Crack Surfaces

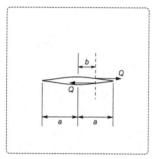

$$K_{II} = \frac{Q}{\sqrt{\pi a}}\sqrt{\frac{a+b}{a-b}} \quad \text{(right crack tip)}$$

$$K_{II} = \frac{Q}{\sqrt{\pi a}}\sqrt{\frac{a-b}{a+b}} \quad \text{(left crack tip)}$$

Single Crack in an Infinite Plate Subjected to a Pair of Concentrated Shear Forces (per unit thickness) on Crack Surfaces

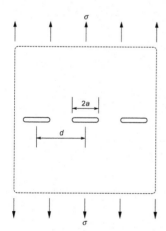

$$K_I = \sigma\sqrt{\pi a}F(\alpha), \quad \alpha = 2a/d$$

$$F(\alpha) = \sqrt{(2/\pi\alpha)\tan(\pi\alpha/2)}$$

Periodic Collinear Cracks in an Infinite Plate Subjected to Tension

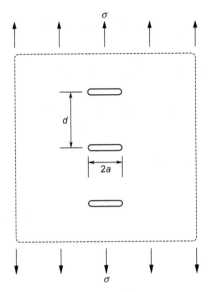

$$K_I = \sigma \sqrt{\pi a} F(\alpha), \quad \alpha = 2a/d$$

$$F(\alpha) = 1 - 0.5(\pi\alpha/2)^2 + 0.375(\pi\alpha/2)^4$$
Accuracy: 5% for $\alpha \leq 0.5$

Periodic Cracks in an Infinite Plate Subjected to Tension

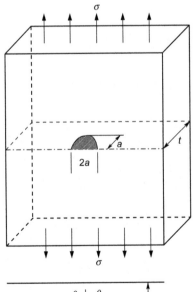

$$K_I = 2\sigma \sqrt{a/\pi} \left[1 + 0.2(2\theta/\pi)^2\right]$$

Accuracy: 3% for $a/h < 0.2$

Semi-circular Crack in an Infinite Plate Subjected to Tension

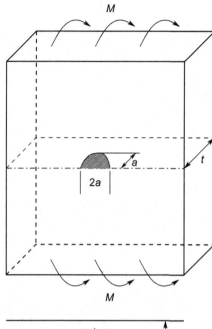

$$K_I = 6.8 \frac{M\sqrt{a}}{h^2} \left[1 - 1.4 \frac{a}{h} + \left(\frac{2\theta}{\pi}\right)^2 \left(0.2 + \frac{a}{h}\right) \right]$$

Accuracy: 3% for $a/h < 0.5$
(M: per unit width)

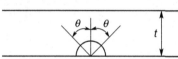

Semi-circular Crack in an Infinite Plate Subjected to Bending

Index

Note: Page number followed by "*t*" indicates tables.